Contents

Introduction
1. Article 555. Planet X and why Physics has failed?
2. Article 501. Planet X induced volcanic eruptions are like an Earth CME
3. Article 514. Stellar Cores are gravitational poles or super protons and white holes
4. Article 523. Planet X and the Solar System: Jupiter and all gas giants are recent acquisitions
5. Article 524. Mercury: created by the Sun due to Planet X
6. Article 526. Planet X and the Moon: the Moon has not always been in sky
7. Article 529. Planet X debris field and water clouds
8. Article 531. Planet X entering earth's atmosphere: creating funnel-shaped clouds
9. Article 532. Planet X or comet reenergizing process
10. Article 533. Planet X induces super nuclear reactions in the earth's core
11. Article 535. The Sun is no longer shining: what has happened to it?
12. Article 534. Planet X in the earth's atmosphere
13. Article 541. Planet X cover-up and planetary formation: where do asteroids come from?
14. Article 542. Planet X observation based energy conversion in the cores of planets and stars
15. Article 558. Astronomical Quantum Mechanics: Gravitational potential and orbitals
16. Article 560. Planet X reveals that planets are smaller versions of their parent stars
17. Article 561. Planet X reveals how the universe began and how it is connected
18. Article 563. Planet X creating sinkholes all over the world
19. Article 566. Planet X creating sinkholes and effects on the human body
20. Article 569. Effect of Planet X Objects as large as the Sun on the Earth: energy levels
21. Article 571. Planet X in our skies: water and vortices
22. Article 573. Planet X reveals why everything rotates and what particles look like
23. Article 576. Planet X larger than the Sun on a collision course with Earth: what happens?
24. Article 577. Planet X created the asteroid belt and rocks in the sky

Books previously published

Book 1: Planet X: the awakening is now.
Book 2: The Planet X Report 2017: Photographic Evidence.
Book 3: Planet X Revealed Gravity and Light.
Book 4: The Sun Simulator
Book 5: Chemtrails: The Silent Killer.
Book 6: Planet X Physicist Articles: Part 1
Book 7: Planet X: The effects on the Earth and the Sun
Book 8: Planet X and the Solar System
Book 9: Planet X and the Hurricane Michael Cover-Up

Introduction

This book details articles in it my recent journey of discovery regarding how the universe works. The understanding grew progressively and was inspired by God the Creator. One day I asked Him what particles look like and I never thought I would get an answer before getting to Heaven but as my understanding grew of the laws governing the universe it suddenly dawned on me what they must look like and how they connect. Since the structure is similar at all levels, the star system structure of a star and planets orbiting is bound to be similar to what we have within particles and indeed we would expect the photon energy to be at the center of the particle as if it is a star inside the particle. There are times however when I don't get it right the first time around or I only get it partially right first and then as I proceed and write another article I realize that there is something more to what I had initially discovered. For example, after seeing how Planet X Objects interacted with earth I assumed that earth would like them to have a limited number of energy layers within but that turned out to be incorrect, the earth as a living celestial objects has all energy levels within down to the densest possible. This is described in the last chapter, chapter 577.

Dr. Claudia Albers

Planet X Physicist

January 28th, 2019

Chapter 1

555. Planet X and why Physics has failed?

Physics is about discovering how the universe works. So physicists are supposed to work to uncover the truth about how the universe operates. However, as time goes on, physics seems less and less able to explain how the universe works and explain what is going on in our Solar System and our planet. I believe that this has occurred because most physicists have failed to properly search for the truth. As a physicist myself, I can place myself within the group of a physicist who, at least initially, failed to search for the truth, simply because I trusted those who were my teachers too much. I thought that if I didn't quite understand something perfectly, after seeking clarification, it was because there was something wrong with my ability to understand, rather with the logic behind the concepts themselves. It was only when I became a teacher and was trying my best to explain everything to my students, in the clearest way possible, that I came across connections that made no sense, and I started to realize that there was something very wrong with physics itself.

Figure 1.1. Students in a lecture.

So why do physicists fail? We fail if we fail to think logically. The main reason why Solar Physicists have failed to uncover the truth about the Sun and what is going on in the Solar System is that, first of all, they fail to realize that nothing happens without a reason. If you park your beautiful undamaged car, in the evening, and in the morning, it has a crumbled back section; do you think that just happened without any cause? Did the car go through some crash cycle? Or did another car crash into your car? Obviously, you think that someone crashed into your car, something happened to cause your car to look the way it now does, and that another vehicle was involved. But Solar Physicists think that the Sun has CMEs, Solar Flares, or may even nova for no reason. It just happens. Because they think it just happens, because they think that there can be an effect without a cause, they have failed to look for the real reason why these things happen and have thus failed to uncover the truth regarding what Planet X is and how the universe really works.

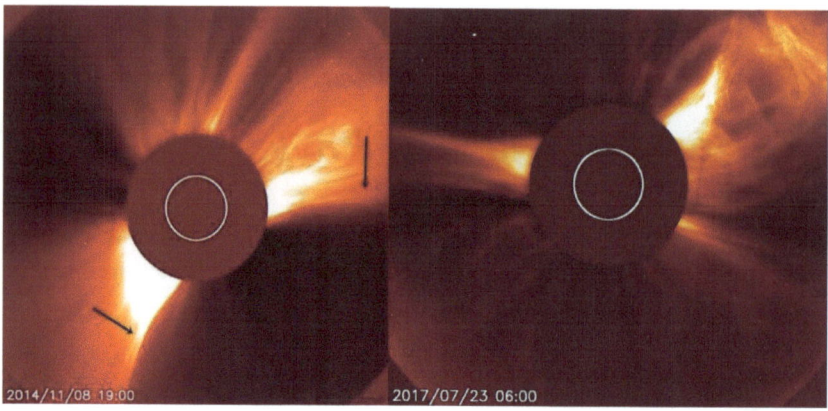

Figure 1.2. Comets (Planet X System Stellar Cores) are the cause behind CMEs. These objects induce matter creation events from the Sun's core, which result in CMEs and Solar Flares (see Article 501: Planet X induced volcanic eruptions are like an Earth CME) [1]. These comets, some of which are much larger than the Sun are the real cause for cataclysmic events in the Solar System, and on Earth; these are the real Planet X.

The other thing that most Solar Physicists, as well as astronomers and astrophysicists, do is ignore observational evidence in favor of theories. They completely ignore images, which clearly show spherical objects in the Sun's corona. Why do they do that? Because, it is impossible, according to theory and they do not want to change theory. And therefore, with the exception of a few individuals, like Halton Arp and James McCanney, they have failed to uncover the truth about how the universe really works. There is of course another motivating factor for steering clear of 'controversial subjects' and agreeing with the accepted theories. If you don't you lose your career or your life. Thus, observational evidence that shows that there was no Big Bang, that redshift is intrinsic and cannot be used to determine distance, and that matter is being created, everywhere in the universe, from within the center of galaxies, to within the center of stars, planets, and moons, is ignored. Matter creation, which is actually the transformation of photon energy (light) into the matter, is how living objects form, they create their own outer layers; they create themselves from the inside out (see Article 522: Stellar Cores are sources of matter or white holes) [2].

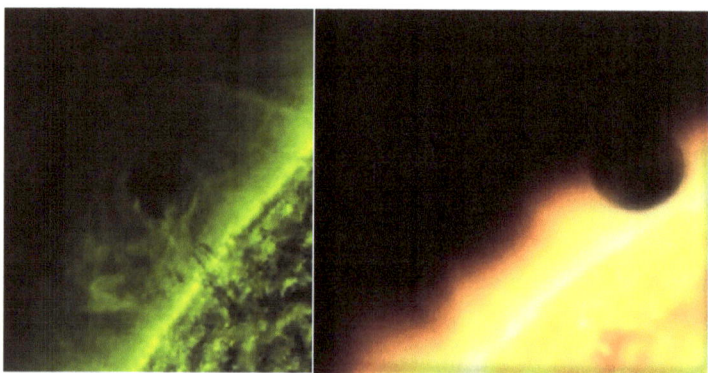

Figure 1.3. Planet X System Stellar Cores or comets in the Sun's corona ignored by accepted physics.

Quantum mechanics has also based on the premise that things happen without a reason, which is illogical. There is a cause for every effect. That is the way it is in our daily life and that is the way we should expect it to be, in the universe. To expect things to be one way, in our daily experience, and to be another way, in the rest of the universe, is illogical. Because what is patently illogical is accepted, quantum theory has failed to understand why changes occur and simply uses broad probabilities to track changing systems. But it is photon energy, within particles, and energy being released from a particular particle, due to one or more lower energy particles, being in its environment which causes changes. In addition to daily life, this principle has also been observed, for centuries, in the form of heat transfer. Why was it not applied to energy transfer, at the quantum level? Because it became expected that logic was not to be applied at that level.

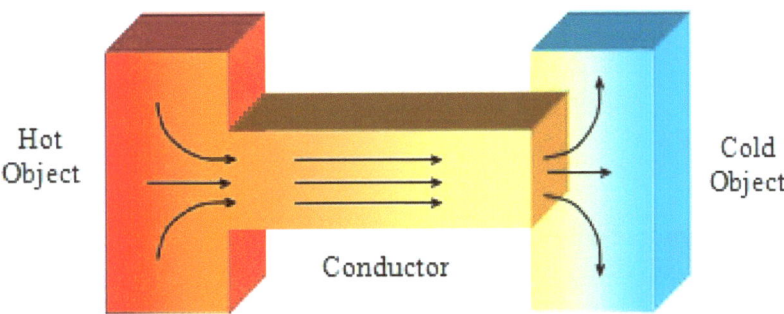

Figure 1.4. Just as heat flows from a hot to a cold object, energy flows from a particle which is lower in energy than other particles, in its environment, and in the same way energy flows from a living celestial object to a Planet X System Stellar Core, because it is lower in energy. The Planet X object would not even be able to come into the atmosphere, of a living celestial object, if it wasn't so low on gravitational (photon) energy that it was no longer able to produce its own outer negative layer. Once in the atmosphere of a living celestial object, if the dead Stellar Core is large enough its electrical influence on the living core creates instability, which results in matter creation within the living core (see Article 533: Planet X induces super nuclear reactions in the earth's core) [3].

It is the absorption or release of photon energy, which results in electrons moving from one energy level to another. This occurs because the interaction strength of the force (charge separation part of the gravitational interaction), which has not yet been discovered by accepted physics, varies according to the amount of photon energy (light) within a particle. Thus, all matter and all energy come from photons, from light (see Book 3: Planet X Revealed Gravity and Light) [4].

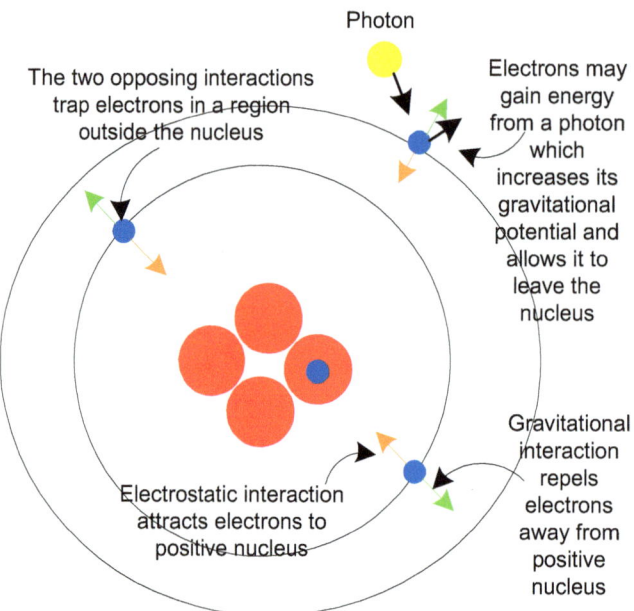

Figure 1.5. Photons are emitted when electrons move to lower energy levels and photons are absorbed when electrons move to higher energy levels. This occurs because the charge-separation force increases when the photon energy, within an electron, increases, and decreases when the photon energy decreases.

Because physicists fail to think logically they somehow never realized that an atom cannot exist unless there is a force to counter the electrostatic attraction between the electrons and the protons, in the nucleus. They have, as a result, also failed to realize that planets have the same structure as atoms and are thus positively charged, on the inside, and have negative outer layers. And what is saddest of all, is that physicists failed to see the real beauty in the universe, where everything from atoms to stars, and galaxies, work according to the same patterns and laws.

References:

[1] Albers, C. (2018). Article 501: Planet X induced volcanic eruptions are like an Earth CME.
[2] Albers, C. (2018). Article 514: Stellar Cores are gravitational poles or super protons and white holes.
[3] Albers, C. (2018). Article 533: Planet X induces super nuclear reactions in the earth's core.
[4] Albers, C. (2018). Book 3: Planet X Revealed Gravity and Light.

Chapter 2

501. Planet X induced volcanic eruptions are like an Earth CME

Planet X System Stellar Cores are entering the earth's atmosphere, just like they entered the Sun's atmosphere or corona, and thus have a gravitational effect upon the earth's atmosphere and surface, which is in the form of a gravitational wave or gravitational vortex. They, therefore, cause the surface of the earth to expand and contract and thus to break up. If the Stellar Core is small and over the ocean, it can create a physical water vortex or waterspout, but even if there is no liquid, which would then give rise to a vortex, the object's gravitational influence still causes a density wave to affect the ground and thus cause it to expand and contract, and break.

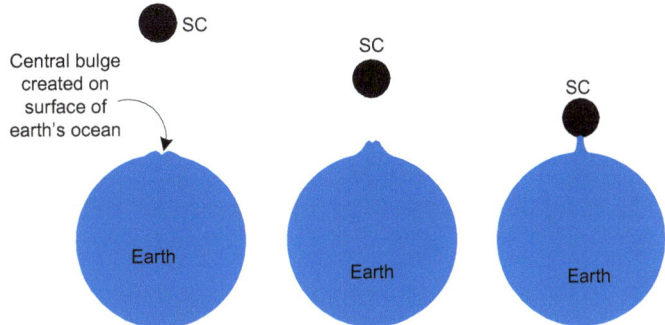

Figure 2.1. As a massive solid object approaches the earth, over its ocean, the gravitational wave, which forms as a result of the interaction between the two objects, causes earth's liquid surface to be deformed so that a wave pattern appears as shown. When the object approaches the earth so that the distance between it and the surface of the ocean is about the same as the wavelength of the gravitational wave, a gravitational vortex forms which connect the two objects.

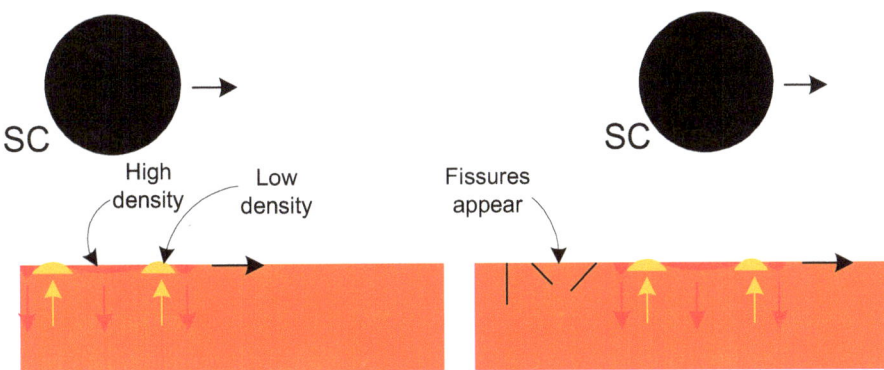

Figure 2.2. A Stellar Core, very close to the earth, is able to generate a tidal force and gravitational waves on the surface of the earth, resulting in alternating regions of higher and lower density, which will cause the ground to break and fissure (see Article 364: Planet X causing the earth's surface to break) [1].

This means that if a Stellar Core passes over a volcano then we can expect the magma inside the volcano to expand and contract and thus likely to be pushed upwards which may then lead to an eruption. However, in a recent eruption of the Popo volcano in Mexico, which erupted on December 16th, 2018, it became obvious that there was an object hovering over the volcano doing its own eruption.

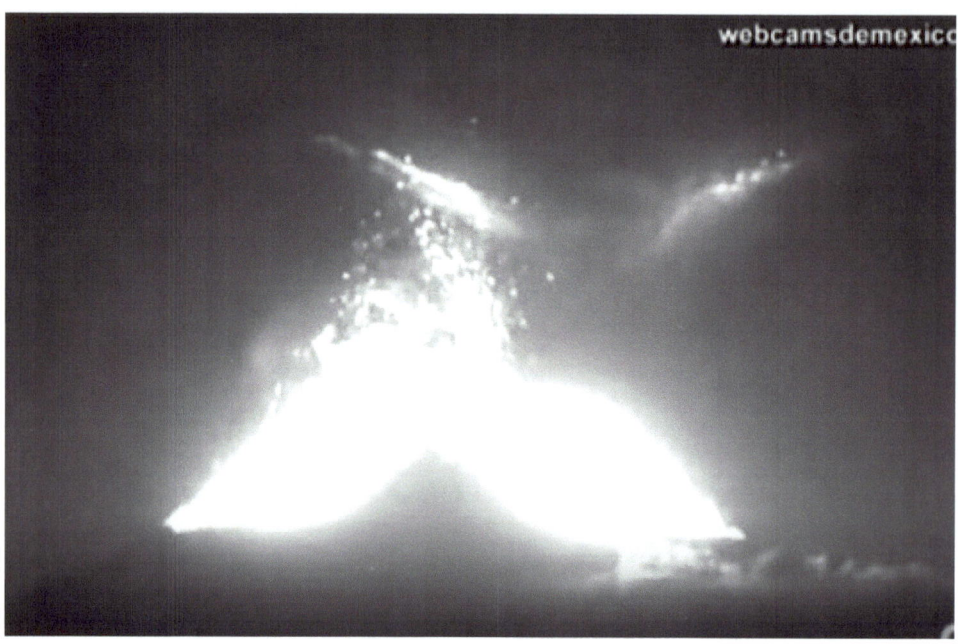

Figure 2.3. Lava seems to be exploding from the volcano, but at the same time, lave seems to be exploding from a point above the volcano. How did exploding lava get to that point in the sky?

Figure 2.4. A view of the volcano during the day shows that there is nothing higher than the volcano around, so the exploding lava could only have been from an object hovering above the volcano, i.e. a Planet X System Stellar Core hovering above the volcano.

Now, where would the lava on the Stellar Core, above the volcano have come from? It most likely got there at the beginning of the eruption. But, since both the volcano and the Stellar Core are seen erupting, at the same time, this suggests they are both reacting to the same effect. Lava comes from deep within the earth and it is very hot and therefore contains a lot of photon energy, and it is likely that the earth responds to the presence of a Stellar Core, which is made of low energy matter, as they are energy depleted objects, by sending high energy matter in the form of magma, in the direction of such an object, if there is a clear path for that high energy matter to travel along, i.e. a volcano, as a volcano provides a connection between the magma within the earth and the surface.

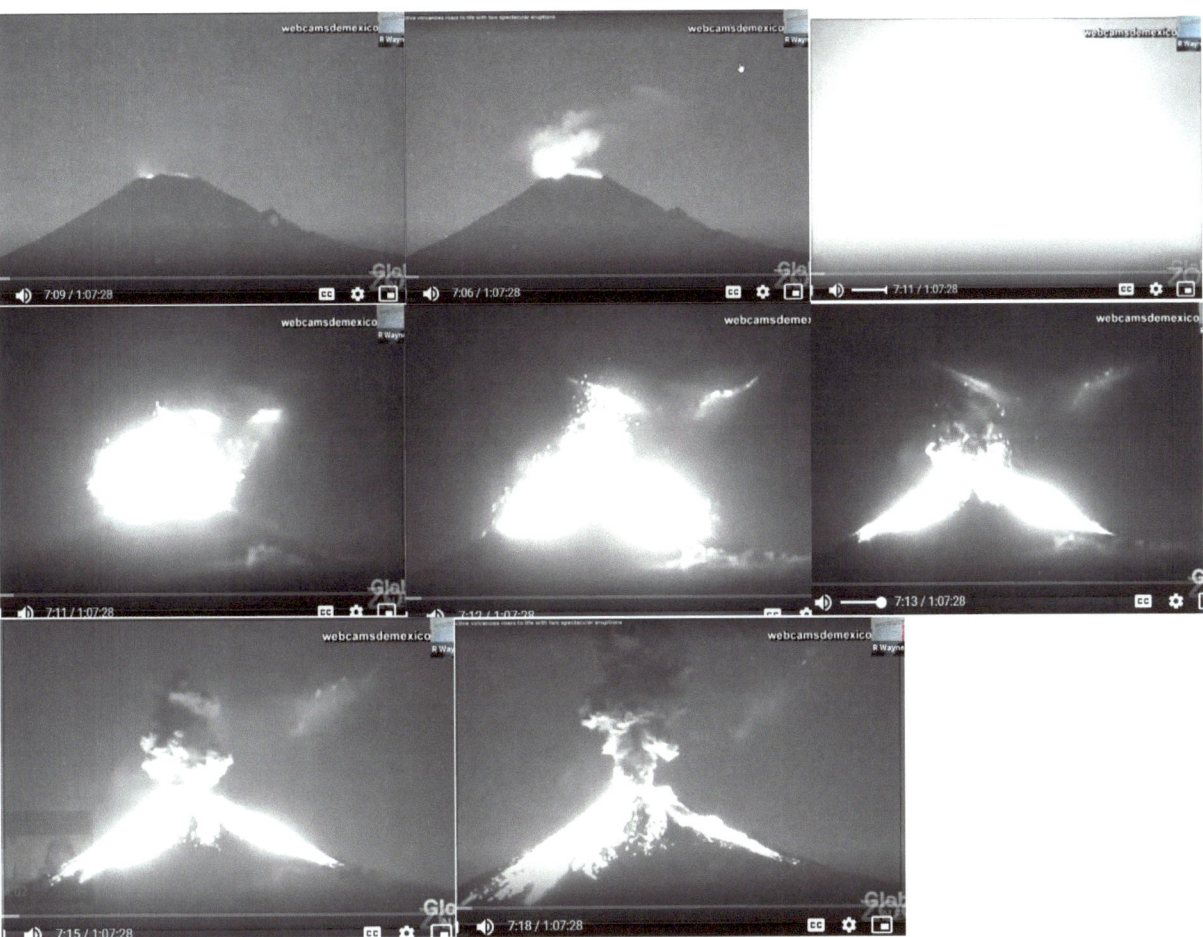

Figure 2.5. Volcanic explosion: Lava starts to issue, then there is a huge explosion of lava but as the brightness decreases it is obvious that there is not that much lava moving down the flanks of the volcano. This was like a light explosion, like a solar flare? As the light decreases it becomes obvious that some of the lava remains suspended in midair and is moving sideways and upwards along what can only be the surface of a Stellar Core, hovering above the volcano, like lava, from the volcano is also seen flowing down along the flanks of the volcano. More lava continues to issue from the volcano and lava seems to also explode and move upwards on the Stellar Core suggesting that both the lava in the volcano and the lava on the Stellar Core are reacting to a reaction from the earth itself. It is possible that the earth's electric field became destabilized in the area, from deep within the Earth; it is as if the Earth is creating a CME (coronal mass ejection on the Sun) and magma is now repulsed away from the earth's core, so that it issues out of the volcano and is repelled away from the center of the earth and hence flows upwards along the surface of the Stellar Core.

In conclusion, Stellar Cores are obviously coming into the earth's atmosphere and are provoking volcanoes to erupt. The eruption could be due to the Stellar Cores' gravitational wave, which would cause magma inside a volcano to expand and contract, and thus become more prone to eruption, but in observing an actual eruption, it looks more like an energy flux, or a flow of matter high in photon energy, flowing toward an object made of matter, which is low in photon energy, and therefore a lot like a CME (Coronal Mass Ejection) on the Sun. The initial eruption even seems to give rise to a light explosion, somewhat like the solar flares seen on the Sun. It seems that both the magma inside the volcano and the magma that has become attached to Stellar Core, continue for the duration of the eruption, to be repelled away from the center of the earth, suggesting that the earth is like a star, on the inside, behaving a lot like the Sun was observed to behave, when approached by Stellar Cores, which resulted in it having solar flares and CME events, during which large amounts of solar plasma exploded away from the Sun.

References:

[1]   Albers, C. (2018). Article 364: Planet X causing the earth's surface to break (in Book 9: Planet X and the Hurricane Michael Cover-Up).

Chapter 3

514. Stellar Cores are gravitational poles or super protons and white holes

Planet X System Stellar Cores are the name I have given to the group of objects, which I started seeing in spacecraft and satellite images of the Sun at the end of 2016. They are able to very closely approach the Sun but never collide with it. They have a weak gravitational influence on other matter. They come in different sizes, only the larger ones are cores of stars; smaller ones, of the size of a small planet and even much smaller than the earth, are obviously the cores of planets or small moons. However, we will continue to use the term Stellar Core to mean the cores of all celestial objects. Planet X System Stellar Cores have low gravitational influence because they are low in gravitational energy, or photon energy (i.e. light), which exists within particles. The light within particles determines the strength of the gravitational interaction (see Book 3: Planet X revealed gravity and light) [1]. Stellar Cores also enter into the earth's atmosphere.

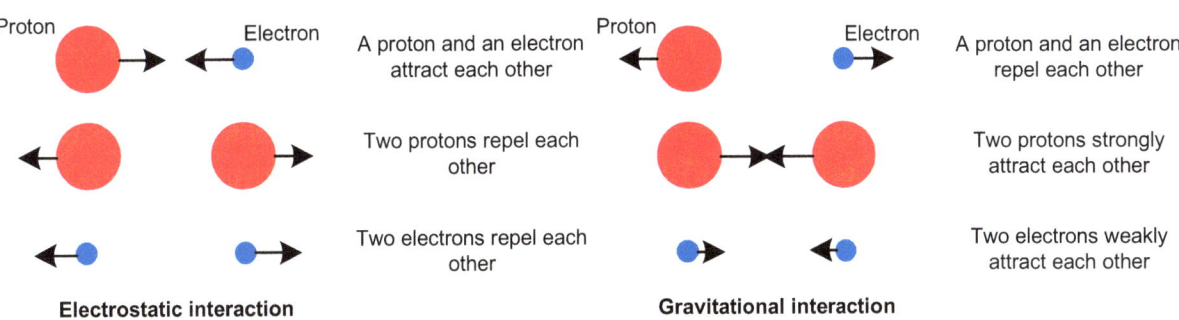

Figure 3.1. The electrostatic and the gravitational interactions between protons and electrons.

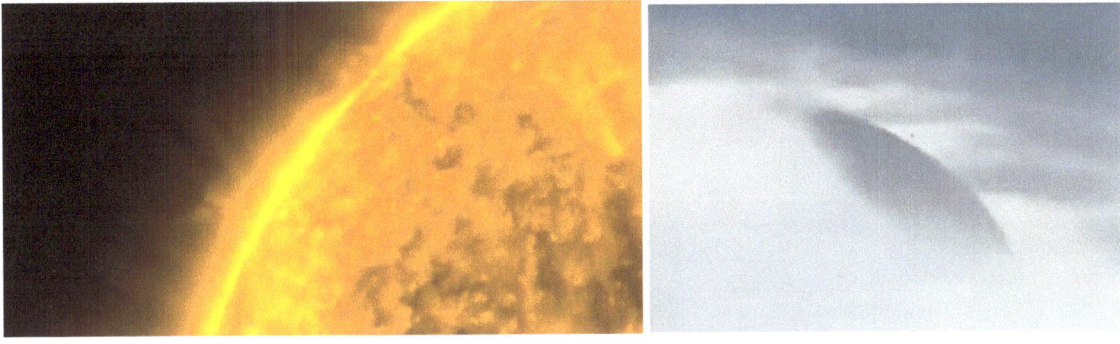

Figure 3.2: SDO image in the 171-angstrom wavelength from October 13th, 2017 showing a dark Stellar Core, which appears to be about half of the size of Jupiter making a gravitational vortex connection with the Sun. **Right:** Photographic evidence of Stellar Cores within the earth's atmosphere: A spherical object immersed in the cloud (Source: R. Wayne Steiger).

These objects enter the earth's atmosphere just like they enter the Sun's corona or atmosphere and interact with the earth in the same way as they were observed interacting with the Sun. They created solar tornadoes, or gravitational vortices with the Sun's surface, as a result of the tidal force they exert on the material at the surface, they are thus able to draw material toward their surface and through that mechanism gain gravitational energy. They do the same with the earth; through the tidal force they exert on earth matter, they are able to create low-pressure weather systems, with cyclonic rotation in the earth's atmosphere. It is, in fact, impossible to have a low-pressure weather system in the earth's atmosphere without these objects being the source of it.

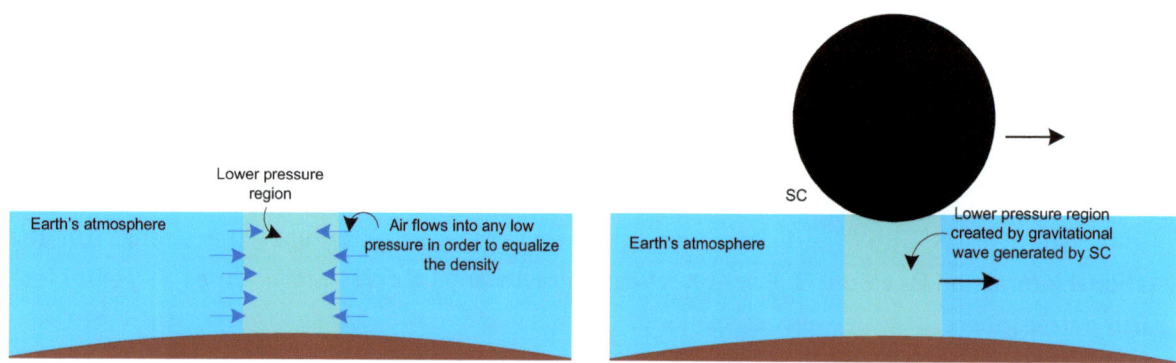

Figure 3.3. Left: A low-pressure region will immediately disappear as air from surrounding region will flow into it in order to equalize the density and therefore the pressure. **Right:** A region of low pressure can only travel in the earth's atmosphere if it exists as a result of the gravitational wave generated by a Stellar Core (SC) inside the Earth's atmosphere.

These objects will also create a gravitational wave on the surface of the ocean, which has a region of low density, just like the low-pressure region they create in the atmosphere, and a region of higher density, which is a region where the gravitational force becomes repulsive and the earth's surface gravitation seems to increase. These regions of gravitational reversal create ocean recession events, which can only be tidal in nature, and which seem to have started occurring in 2017, when, Stellar Cores capable of creating larger gravitational vortices capable of influencing the ocean in this manner, started entering the earth's atmosphere.

Figure 3.4. The ocean receded in August 2017 from the coast in Uruguay and Brazil.

The larger the area affected by the gravitational wave or vortex, which they create on the ocean surface, the larger the object giving rise to it, is likely to be. These objects are always accompanied by a huge debris field.

Figure 3.5. This Hi1 A image of the object shows that it is surrounded and followed by a huge cloud of debris.

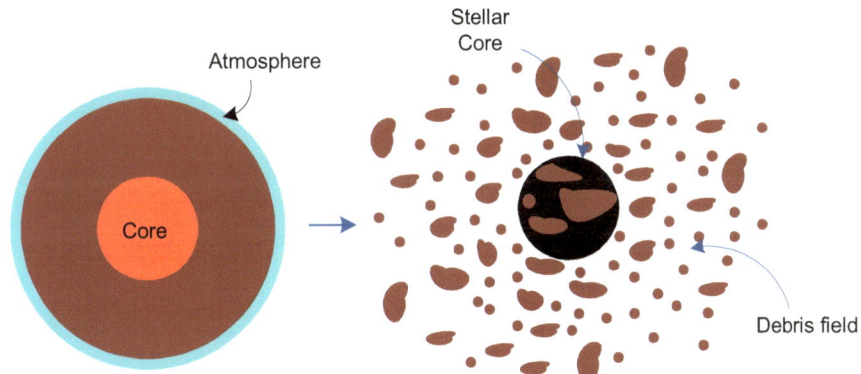

Figure 3.6. A living celestial object turns into a Stellar Core when it runs out of energy. A Stellar Core is made up of its original core and will be surrounded in a debris field. The debris field is the material that was once part of the object's outer layers of material. Water which was a part of the planet turns into tiny droplets of water as a result of the loss of gravitational energy and envelops the debris pieces and the Stellar Core, i.e. the dead core of the celestial object.

Figure 3.7. A large bank of cloud advances along the ground toward a city in China. The cloud was fog, pink Planet X fog, not a sandstorm as had been reported as it was clearly not made of sand as sand swirls up and down. This was cloud and thus tiny droplets of liquid water and it was emitting pink light and since it is impossible for water to remain suspended in the atmosphere as it is denser than air this is not earth water this is Planet X debris water.

There are also large pieces of the broken planet in the debris field which will enter the atmosphere when the core approaches earth or enters the atmosphere. The larger the pieces the longer it will take to absorb enough energy to reach the surface and may thus remain suspended, in the atmosphere, for a long time. Stellar Cores themselves never touch the surface and I have said that this is due to the fact that they are positively charged, which is true. But it is more than that. Stellar Cores are not like normal matter, they are sources of the gravitational field, they are gravitational poles, and like protons within the nucleus. They can, therefore, be called super proton. Protons attract each other very strongly within the nucleus but never collide, the force of attraction is extremely strong and is thus called the strong nuclear force, but it becomes repulsive once the protons approach each other beyond a certain limit, and the repulsive force very quickly tends to infinity, at closer distances, so that protons cannot collide. This occurs because both approaching particles are positively charged and the electric force becomes infinitely repulsive at such close distances.

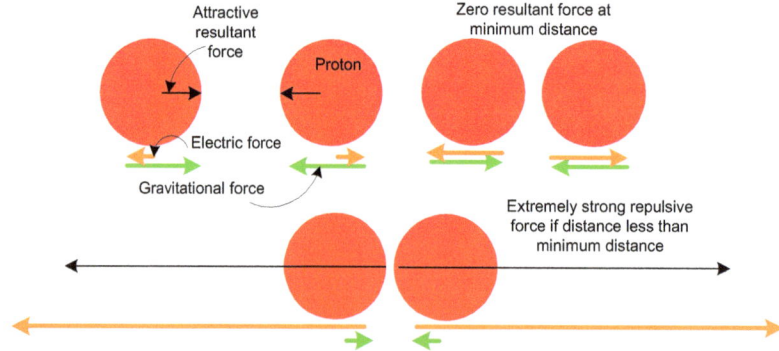

Figure 3.8. Protons are strongly attracted to each other at close distances but there is a minimum distance at which the strong (gravitational force) is the same as the electric, and beyond that distance the electric repulsive force increases extremely quickly, tending to infinity so that protons can never

collide. The same occurs with Stellar Cores. Stellar Cores are sources of the gravitational field and positive electric field and are thus like super proton.

Matter, which is not a part of a Stellar Core, is different from matter making up a Stellar Core; it responds to the presence of a Stellar Core and is attracted to it. This matter behaves like a fluid, if it is in small pieces, in comparison to the size of the Stellar Core, and develops wave-like effects to the point that gravitational vortices form, which are actually gravitational diffraction effects so that in the same way that a small slit leads to light diffraction as the slit acts as a source of light, a hole in a sink acts as a source of gravity. Gravitational diffraction manifests in the form of a vortex. Water in a sink, going down a hole, creates a vortex, which is a diffraction effect, due to water being acted upon by gravity, through a small hole, which acts as a source of gravity.

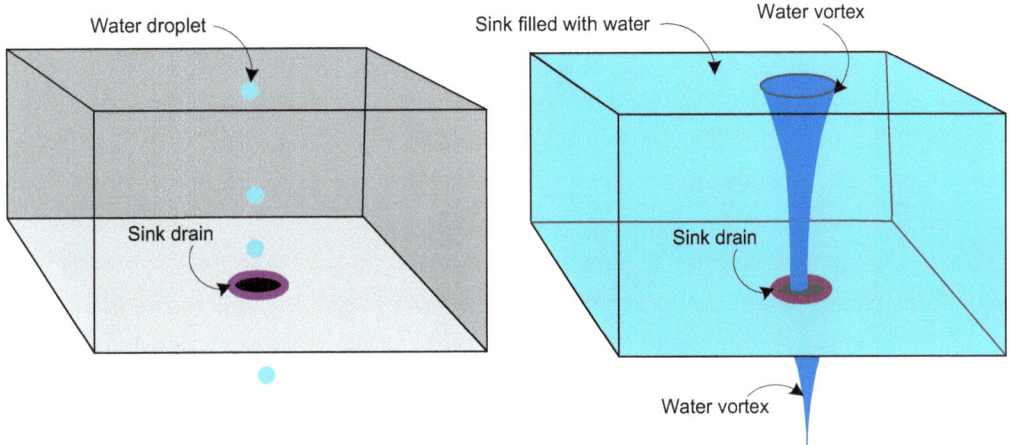

Figure 3.9. A water droplet will fall right through a sink drain without any gravitational vortex effects appearing, i.e. no gravitational diffraction because the hole is large in comparison with the amount of material falling through it. But if the sink is filled with water, the vortex, or the diffraction effect, appears. The vortex is widest where the force is strongest and in the case of the sink, it is strongest at the surface of the water. The water going through the hole continues to fall as a vortex, after it passes the hole, indicating that diffraction occurs on both sides of the hole.

Light in the form of free photons acts in a similar manner when forced to move through a small hole or slit, the slit acts as a source of light and a diffraction pattern arises.

Thus, matter which is not a part of a Stellar Core acts like electrons in a nucleus, they respond to the attraction of a nucleus and gather around it. In this way a celestial object is like an atom and may be called a superatom, and just as an atom, it has several energy levels, or layers, starting from the magma layer down up to the atmospheric layer which has the lowest gravitational potential (gravitational energy per unit mass).

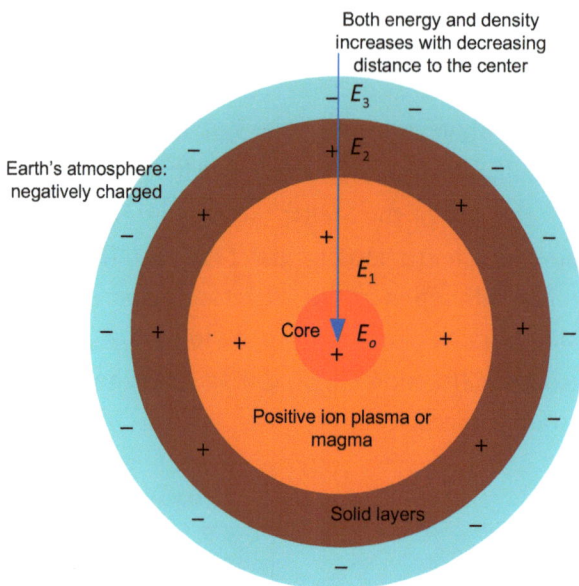

Figure 3.10. Gravitational energy increases with depth and is in the form of energy levels, matter in the core has the most energy, matter in the molten rock region is in the next energy level, matter in the solid rock layer is lower in gravitational energy, and matter in the atmosphere is in the last and lowest energy level. Density also decreases, with distance, from the center; the Stellar Core is the densest part of the superatom, and the outer gaseous layer has the lowest density.

Matter is rich in protons up to the surface but the atmosphere contains more electrons than protons. The surface matter is neutral and will thus contain as many protons as electrons.

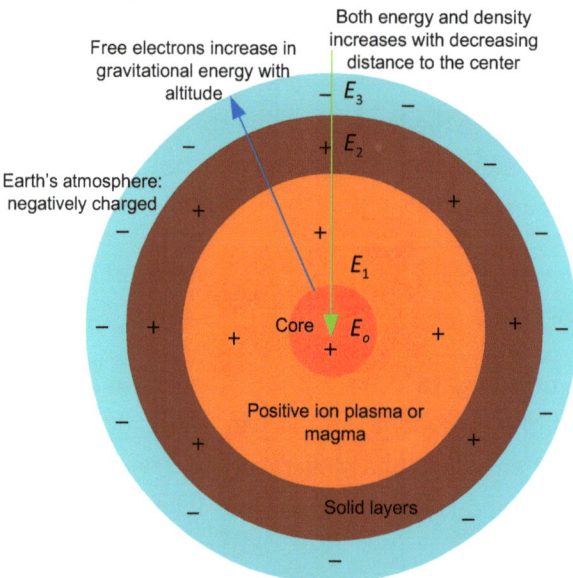

Figure 3.11. Electrons have negative gravitational energy because they are negatively charged and therefore free electrons with the most gravitational energy occur at the top of the atmosphere.

16

Stellar Cores vary in size and most likely in gravitational energy as well. The smallest and weakest are only able to draw material from the earth's atmosphere and thus material which belongs to the earth's lowest energy level, stronger Stellar Cores will be able to draw water from the surface and possibly solid matter such as lose dust and rocks. The largest and strongest will be able to draw matter from the highest energy level of drawable matter, i.e. from the magma part of the earth. Stellar Cores which draw this highest gravitational energy matter seem not to attract matter from the other levels, it is as if they differentiate between energy levels and only draw from the energy level right below their own. For example, some Stellar Cores were observed only drawing atmosphere from the Sun, whilst others did not draw atmosphere but drew denser matter from the Sun's liquid surface or chromosphere. Whilst even larger, and stronger, Stellar Cores, drew matter from so close to the Sun's core that a huge quantity of plasma often erupted towards them, from way inside the Sun, thus creating huge CMEs.

Figure 3.12. Left: Stellar Core in the Sun's corona attract solar atmosphere. **Center:** Stellar Core attracts liquid material from chromosphere, it does not attract atmosphere from the Sun but another object higher up seems to be covered in coronal or solar atmosphere plasma. **Right:** Huge Stellar Core causes huge explosive ejection of chromospheric plasma which must have come from deep within the Sun.

Figure 3.13. Left: What seems to, at first, to be a break in the clouds is convex, and thus a spherical object, poking through the cloud, and therefore a Stellar Core, in the earth's atmosphere. It is surrounded by a pink and grey cloud, which would have formed around dust, and which would be a part of the object's debris field. The object seems to be drawing atmosphere and is thus a weak Stellar Core. **Right:** A water spout being drawn upwards is a gravitational vortex indicating a Stellar Core (source of gravitational force) is within the cloud and drawing water upwards. This Stellar Core is stronger, i.e. it

has a higher gravitational potential than the Stellar Core on the left, because it is able to draw denser matter, i.e. water, from the earth's surface.

Thus, the largest and strongest Stellar Cores will draw plasma from deep within the earth and will not affect the atmosphere or the ocean. They will, however, cause volcanic eruptions and earthquakes

Figure 3.14. A volcanic eruption, like a CME on the Sun, is induced by a large Stellar Core which has a comparable electrical potential to a magma layer inside the earth and induces that layer into having a matter creation event.

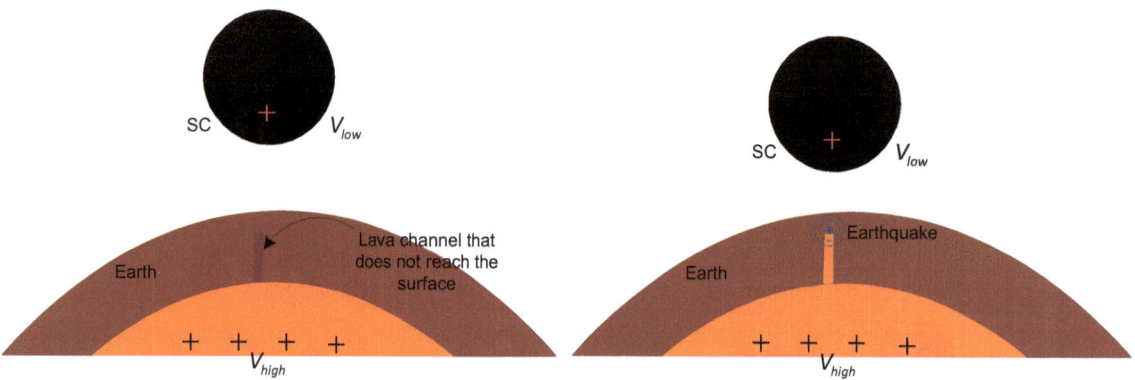

Figure 3.15. An earthquake is essentially the same as a volcanic eruption and is due to a large Stellar Core causing a matter creation event which ejects plasma outwards but if it finds no outlet through a volcano it causes an explosion inside the earth which reverberates as an earthquake.

Stellar Cores will form as a result of the electric interaction when a star or galactic nucleus goes through a matter creation event, in which very high energy photons move through a very high electric field, which causes matter to appear from within the photons. The matter appears to be fluid at first but it condenses into a celestial object with a core, i.e. a Stellar Core or a source of the gravitational field. This is what Halton Arp observed and described was occurring in the universe, in his book 'Seeing Red'; active galaxies were ejecting matter which was condensing into quasars, which eventually turned into galaxies [2]. This indicates that the cores of planets and stars are sources of matter and thus white holes, the opposite of black holes which supposedly draw all matter around them inwards.

In conclusion, Planet X System Stellar Cores are sources of the gravitational field and act as super proton, all celestial objects have a Stellar Core at its center. Matter which is not a part of a Stellar Core responds to it and behaves as matter in different energy levels outside the nucleus of an atom, i.e. like electrons. A Stellar Core attracts matter which is close to it, in gravitational potential and does not attract matter that is much lower than it, in gravitational potential. Only the largest and strongest Stellar Cores produce CMEs and volcanic eruptions or earthquakes.

References:

[1] Albers, C. and C'one, S. (2018). Book 3: Planet X revealed gravity and light.
[2] Arp, Halton (1998). *Seeing Red*. Apeiron, Montreal.

Chapter 4

523. Planet X and the Solar System: Jupiter and all gas giants are recent acquisitions

Planet X System Stellar Cores have been coming into the Solar System as comets for thousands of years as the first documented observation of a comet occurred about 3000 years ago. The Planet X System surrounds the Solar System as it fills the region of space called the Interstellar Medium (see Article 339: Planet X and the Interstellar Medium: can we leave the Solar System) [1] and others refer to as the Oort Cloud, where comets come from. Comets are not small as we have been led to believe, many are much larger than the Sun. The Planet X System is made up of thousands of dead star systems comprising of a star and planets, which have now turned into a dead core surrounded by a debris field. I use the term Stellar Core to describe the cores of all stars and planets. The Planet X System Stellar Cores are all dead in that they are photon gravitational energy depleted and as a result have a very low gravitational influence (see Article 210: Stellar Core gravity: tidal and G is not constant) [2] and cannot create matter, which all living Stellar Cores seem to be able to do which makes the term 'white hole' an appropriate descriptive name for them (see Article 514: Stellar Cores are gravitational poles or super proton and white holes) [3].

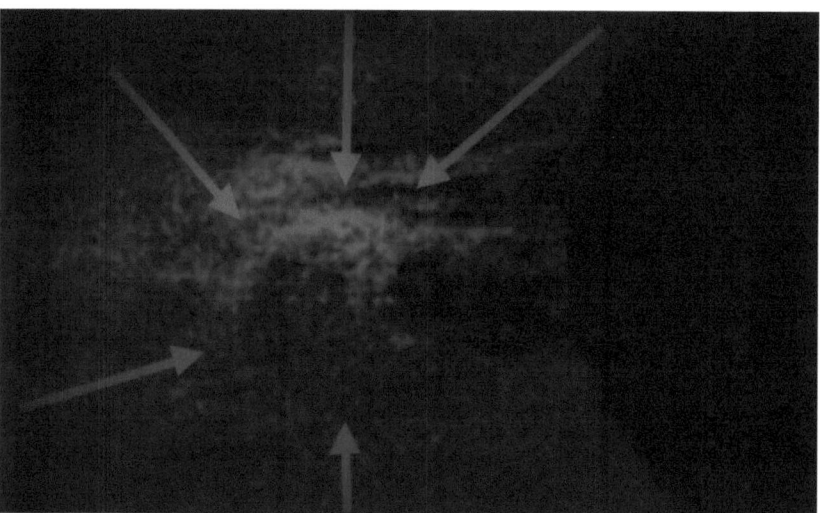

Figure 4.1. Stellar Core close to the Sun in a LASCO C2 image, CME plasma (solar chromospheric plasma or solar magma) can be seen going outwards from the Sun towards the object. A CME event is a matter creation event.

Thus, Stellar Cores are not able to create matter but they are able to induce the cores of living celestial objects to create matter, which they then absorb as an outer layer. They are however only able to absorb matter of comparable gravitational energy so the larger Stellar Cores go to the Sun to obtain their new outer layer and the smaller ones will go to other planets in the Solar System that have living cores in them, such as the Earth (see Article 576: Planet X larger than the Sun on collision course with

Earth: what happens?) [4]. Thus, dead Stellar Cores, which were once living stars would most likely go to the Sun and dead Stellar Cores which were once living planets will go to living Solar System planets. Since they can only acquire material from a Solar System object, with material of equivalent gravitational potential, to what the dead's Stellar Core's gravitational potential current is, their overall gravitational potential cannot change and they will thus continue to exist as objects of much lower gravitational potential than they were as a living celestial object. Thus, objects that were once small stars will now only have the gravitational potential of small planets and very large stars will now have the gravitational potential of large planets.

Planet X System Stellar Cores are observed being ejected from the Sun after a CME, or matter creation event, which suggests that they are repelled away from the Sun, once they have gathered some solar material. They may then be ejected out of the Solar System or they may go into orbit around the Sun. If they go into orbit their acquired orbital position would be associated to their gravitational potential, with the highest gravitational potential objects acquiring orbits close to the Sun and lower gravitational potential objects acquiring orbits further from the Sun. This is because of gravitational potential increases toward the center of a planet or decrease from the center of the Sun, outwards.

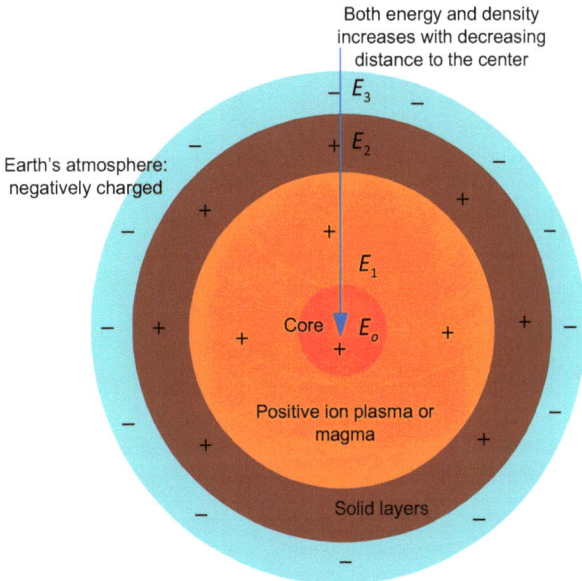

Figure 4.2. Gravitational energy per unit mass, or potential, increases with depth and is in the form of energy levels, matter in the core has the highest potential, matter in the molten rock region is in the next energy level, matter in the solid rock layer is lower in gravitational potential, and matter in the atmosphere is in the last and lowest energy level (see Article 514: Stellar Cores are gravitational poles or super proton) [1].

This would mean that planets with orbital radii, greater than earth's orbital radius, will have a lower gravitational potential than earth's gravitational potential. Thus, since Jupiter orbits in the outer Solar System, Jupiter's core will have a lower gravitational potential than earth's core. But Jupiter is 11 times larger than Earth. The earth's core is now 1/5th (20%) the radius of the earth, but we know that the earth expanded to about twice its initial radius, so the earth's core was initially 40% of earth's radius, which would make that a likely estimate for all living celestial objects (see Article 521: Planet X and the Flood: the earth expanded due to matter creation by its core) [5]. Thus, if Jupiter was a living celestial object its core should be 40% of Jupiter's radius. Therefore, Jupiter's core should be 22 times larger than earth's core, i.e $r_{JCore} = 22 r_{EC}$

Jupiter should, therefore, have a much greater gravitational potential than the earth. The largest cores would logically contain more energy and be able to generate more matter, which is why stars are larger than planets and galactic nuclei are larger than stars. Thus, Jupiter cannot be a living celestial object, it must have a dead core and thus Jupiter must be a Planet X System Stellar Core, which became a Solar System object.

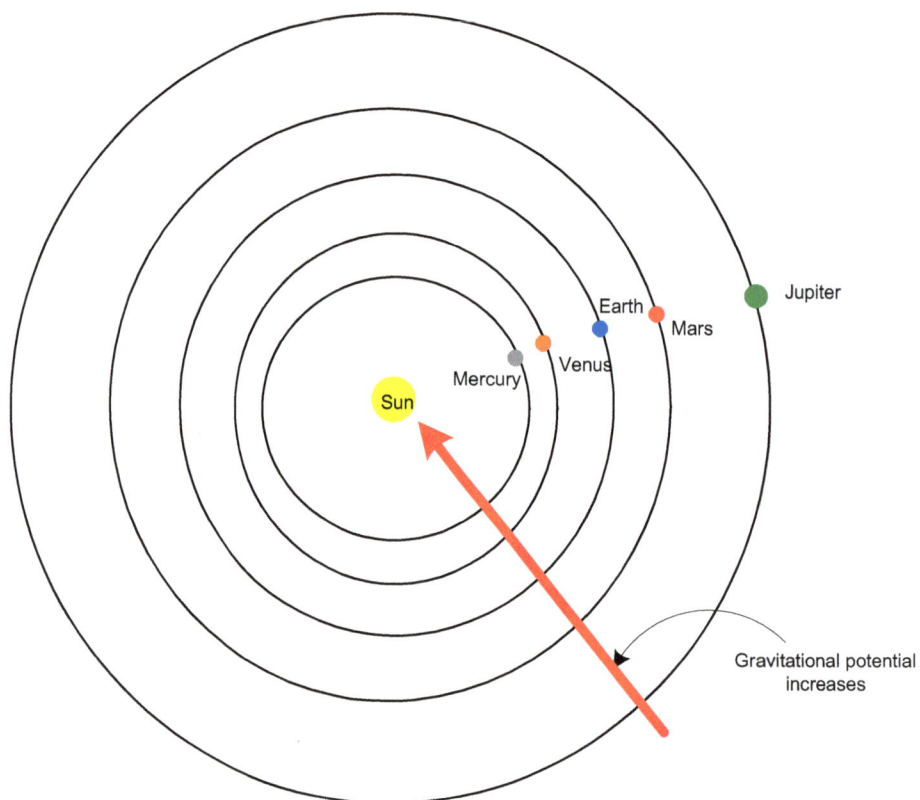

Figure 4.3. Planets orbiting further out from the Sun have a lower gravitational potential but larger objects should have larger cores and larger cores, if living, should have greater gravitational potential. This means that Jupiter, as the largest Solar System planet, cannot be a living planet; it must have a dead core and is thus a Planet X System object which became a Solar System planet.

This means that Planet X System Stellar Cores, once they have been able to acquire an outer layer of material, go into orbit in the Solar System, and thus become acquired planets in the Solar System. This also means that all the gas giant planets, since they are all much larger than the earth must be acquired Solar System planets, which came in as Planet X System Stellar Cores. Since Mercury and Venus, are smaller than the earth, and orbit in the inner Solar System, they are, most likely, living planets with larger cores than earth's core. Mars is smaller than Earth and has volcanoes on its surface, so it must also be a living planet with little less gravitational potential than the earth, which has also been affected in a similar manner to earth, by the Planet X System Stellar Cores, i.e. at least one object approached it and induced it to have matter creation events, which then caused volcanoes to form, as the planet ejected magma. This also most likely caused Mars to expand, just like the earth expanded, as a result of the encounter (see Article 521: Planet X and the Flood: the earth expanded due to matter creation by its core) [5].

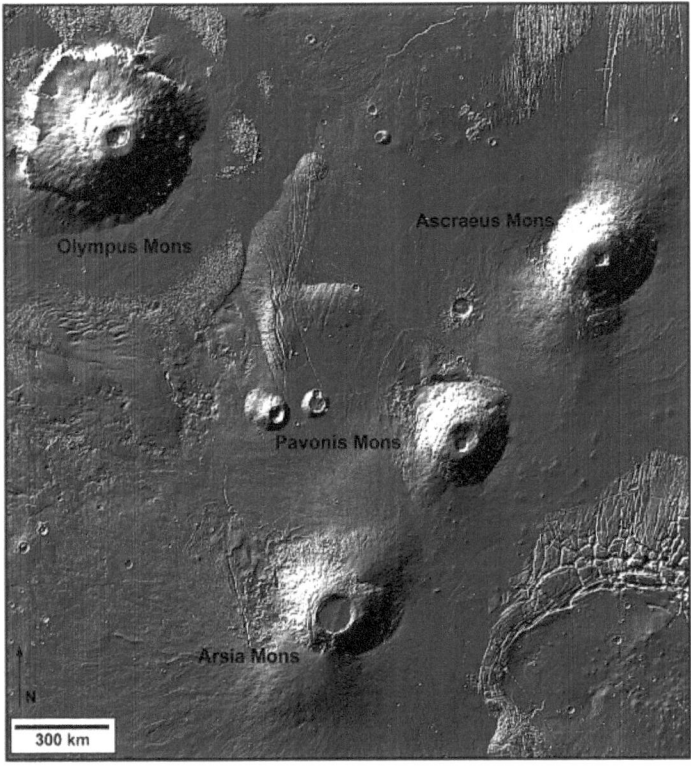

Figure 4.4. Volcanoes on Mars indicate that it is a living planet which has been affected by the Planet X System in the past. A Planet X System Stellar Core must have come very close to it and caused it to undergo a mass creation event which led to the formation of volcanoes on its surface.

Since the Planet X System seems to have started affecting the Solar System in the last few thousand years, the gas giant planets must have been acquired in the last few thousand years and possibly at the time of the Great Flood, when the earth seems to have gone through a huge cataclysmic event, as a result of perhaps the first Planet X System objects entering the Solar.

In conclusion, Planet X System Stellar Cores, which come into the Solar System, seems to remain in the Solar System and eventually settle into regular orbits, after acquiring an outer layer of material from living Solar System planets. Jupiter and all the gas giant planets, because they are much larger than earth, but yet have orbital radii associated with a much lower gravitational potential than earth's gravitational potential, must have dead cores and must, therefore, be dead planets which have been acquired as planets in the last few thousand years.

References:

[1] Albers, C. (2018). Article 339: Planet X and the Interstellar Medium: can we leave the Solar System (see Book 8: Planet X and the Solar System).

[2] Albers, C. (2018). Article 210: Stellar Core gravity: tidal and G is not constant (see Book 6: Planet X Physicist Articles Part 1).

[3] Albers, C. (2018). Article 514: Stellar Cores are gravitational poles or super proton and white holes

[4] Albers, C. (2018). Article 576: Planet X larger than the Sun on a collision course with Earth: what happens?

[5] Albers, C. (2018). Article 521: Planet X and the Flood: the earth expanded due to matter creation by its core.

Chapter 5

524. Mercury: created by the Sun due to Planet X

Figure 5.1 shows a Stereo COR2 image of the Sun in which several objects that should not be there can be seen. Figure 2 shows an enhanced version of the bottom part of the image which gives a better idea of what these objects are. They are clearly spherical and can only be Planet X System Stellar Cores as large numbers of these objects have been observed in solar observing spacecraft in recent years, usually in the Sun's corona. In this image, it is however not clear whether the objects are very close to the Sun or not.

Figure 5.1. Stereo A COR2 image of the Sun's outer corona showing several objects that are not supposed to be there. The largest object is spherical and seems to be draped in light-emitting plasma. The object has a large apparent size and is not likely to be under the Sun suggesting that the light coming from it cannot be reflected light, but that the object is emitting light from the material that its northern hemisphere seems to be covered in. Since the covering seems to be spherical in shape the object is likely to be spherical.

The largest object, in the above image, appears to have its northern hemisphere covered in a smooth material that emits light. This is likely to be chromospheric magma from the Sun or another Solar System object, which has a living core, since it has become clear that all living celestial objects have the capability to create matter from their cores, and that the material created is in the form of hot liquid plasma or what we call magma, in the case of earth. But even active galactic nuclei create matter in the form of liquid plasma, which then condenses into a quasar, according to Halton Arp, in his book 'Seeing

Red' [1]. But not all magma will have the same gravitational potential, magma created by the Sun will have much more than magma created by the earth and magma created by a galactic nucleus will have much more than a star's; thus, Planet X System Stellar Cores can obtain magma at different levels of gravitational potential by going to different objects in the Solar System capable of creating it (see Article 514: Stellar Cores are gravitational poles or super proton and white holes and Article 523: Planet X and the Solar System: Jupiter and all gas giants are recent acquisitions) [2, 3].

Figure 5.2. Stereo COR2 image shown in figure 1 now enhanced (by Scott C'one) where we can see that the object on the right has a circular outline and the larger object has a circular outline which seems to be around and behind the object and may, therefore, be its debris field. The brighter part of the object on the right is most likely also acquired liquid plasma which both objects most likely obtained from indicating a matter creation event on one of the Solar System objects, which suggests that they are most likely large Planet X objects.

Now, in figure 1, the brightest object at the top, which was in the form of a cross, was Mercury and behind Mercury, we could see a large object. This object had a width which was 1/3 the size of the Sun's width as indicated by the white circle on the occulter. Since the object was behind Mercury and Mercury was at this time aligned with the Stereo A spacecraft, on the other side of the Sun and since Mercury's average orbital distance is 0.4 au, the object was thus 40% further from the spacecraft than the Sun, which would make the object's radius $0.42r_S$, where r_S is the Sun's radius. Since Mercury has a radius of 2440 km and the Sun's radius is 695 508 km, this object would be about 120 times larger than Mercury. This is absolutely huge, and since Planet X System Stellar Cores usually make Solar System objects of comparable gravitational potential their host it is unlikely that such a large object has made Mercury its host, but yet it is not impossible as Mercury, which is the closest planet to the Sun, must also be the planet with the largest core and the planet with the highest gravitational potential. This means that Mercury must be almost entirely a core and must also have a very thin solid crust. If Mercury is hosting such an object then it will, almost certainly, also be expanding just as earth expanded at the time of the Great Flood (see Article 521: Planet X and the Flood: the earth expanded due to matter creation by its core) [4].

Figure 5.3. The object behind Mercury appears to have one third the width of the Sun which would make it 120 times larger than Mercury.

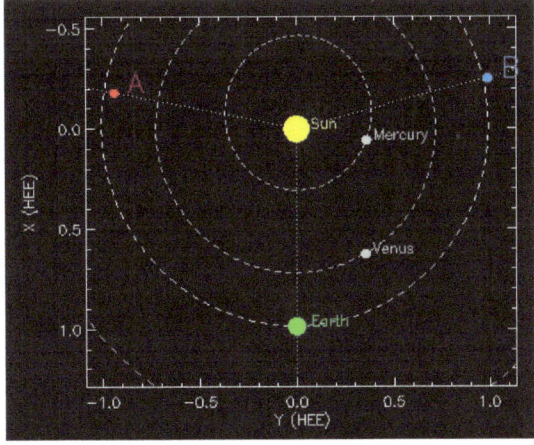

Figure 5.4. Stereo A's position indicates the object seen in the Stereo A COR2 image above the Sun is in the right position to be Mercury.

Now Mercury is 60% closer to the Sun than Earth and therefore its core is likely to be 60% larger than earths. Earth's core has a radius of 1220 km (760 miles), so Mercury's core is likely to be about 2000 km in radius or about 80% of Mercury's radius. A research project done with radar suggests that the liquid layer goes up to 85% of Mercury's radius, which would mean that its solid crust is only 15% of its radius or 366 km (227 miles). This would make Mercury's magma layer only 5% of its radius. A planet with such a large core and such a small magma layer suggests that it was not long ago that the planet came into being as we would expect a planet to grow its own magma layer from its core. This is what seems to occur with galaxies, galactic cores eject galactic liquid plasma or galactic magma which condenses into a quasar or small galactic core which then starts ejecting its own magma which condenses into star clusters that then arrange themselves into spirals, or ring structures, i.e. the arms of the galaxy [4]. Thus, a planet would be ejected from its parent star, as a core, which then ejects magma, which forms into a liquid inner layer followed by a solid outer layer where the liquid layer solidifies; it will most likely also slowly form an atmosphere. The atmosphere will separate from the magma which the core creates and thus a planet slowly forms (see Article 514: Stellar Cores are gravitational poles or super proton and white holes) [1].

Figure 5.5. Mercury's closeness to the Sun suggests that it has a very large core which then suggests that it is a newly formed planet, ejected as a planetary core by the Sun a few thousand years ago.

So the fact that Mercury is not much larger than its core suggests that it is a newly formed planet, which was ejected as a core by the Sun not too long ago. Since we are dealing with astronomical time scales though, this would most likely mean that Mercury was spat out by the Sun a few thousand years ago. This places the creation of Mercury by the Sun at about the same time as the Great Flood, when the Planet X System seems to have started affecting the Solar System. This suggests that it was the interaction between the Sun and the Planet X System objects that came in at the time that induced the Sun into such a frenzy of matter creation activity that it created a new planet.

In conclusion, a Stereo A COR2 image seems to show an object which is 120 times larger than Mercury behind it. If such an object is close to Mercury, the planet is most likely having violent volcanic eruptions and growing in size as the earth seems to have done during the period of the Great Flood. As Mercury is so close to the Sun it will have a very large core which suggests that it is a newly formed planet, most likely ejected as a planetary core by the Sun a few thousand years ago when the Planet X System objects which came in at the time induced it into intense matter creation events.

References:

[1]   Arp, Halton (1998). *Seeing Red*. Apeiron, Montreal.
[2]   Albers, C. (2018). Article 514: Stellar Cores are gravitational poles or super proton and white holes.
[3]   Albers, C. (2018). Article 523: Planet X and the Solar System: Jupiter and all gas giants are recent acquisitions.
[5]   Albers, C. (2018). Article 521: Planet X and the Flood: the earth expanded due to matter creation by its core.

Chapter 6

526. Planet X and the Moon: the Moon has not always been in the sky

It is widely believed that the moon has been with the earth for millions of years. However, the Planet X System, which could also be called the Destroyer System, of dead stars and planets, has been entering and affecting the Solar System and our planet, for thousands of years (see Article 523: Planet X and the Solar System: Jupiter and all gas giants are recent acquisition) [1]. These objects are depleted in energy and have a low gravitational influence, and are also not able to create matter as living cores, of stars and planets, are (see Article 514: Stellar Cores are gravitational poles or super proton and white holes) [2].

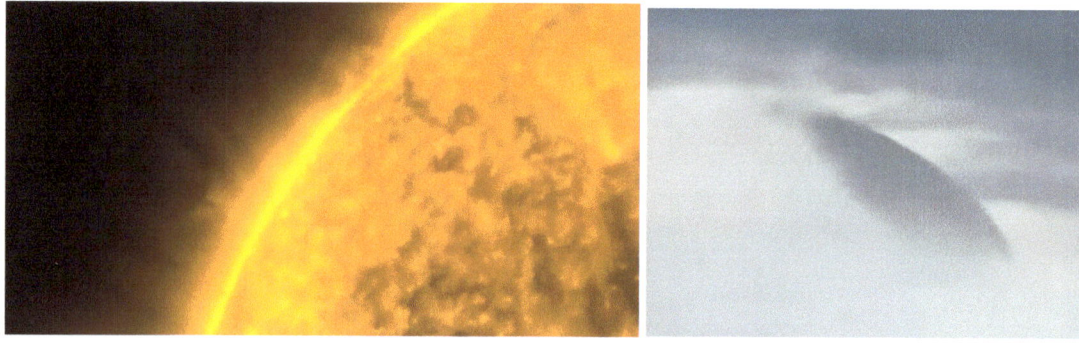

Figure 6.1: SDO image in the 171-Angstrom wavelength from October 13th, 2017 showing a dark Stellar Core, which appears to be about half of the size of Jupiter making a gravitational vortex connection with the Sun. **Right:** Photographic evidence of Stellar Cores within the earth's atmosphere: A spherical object immersed in the cloud (Source: R. Wayne Steiger) It is because they are always immersed in clouds that they are so difficult to observe within the earth's atmosphere.

Thus, Planet X System Stellar Cores, because they are depleted in gravitational energy, cannot create matter. This is because matter creation arises as a result of light or photons being transformed into matter (see Book 3: Planet X Reveals Gravity and Light) [3]. Energy conservation is not violated. But even though they can no longer be a source of matter Planet X Objects will still retain a strong positive electric field and by closely approaching a living celestial object, with a living core, they are able to induce it into having matter creation events, or transforming some of their gravitational photon energy into matter. This is what causes the Sun to have a CME, and for planets with rocky surfaces, like the earth, it results in volcanic eruptions, as the matter created is in the form of positively charged liquid plasma, i.e. magma. Earthquakes are also the result of matter creation events but, in this case, the magma was not able to reach the surface of the planet (see Article 501: Planet X induced volcanic eruptions are like an Earth CME) [4].

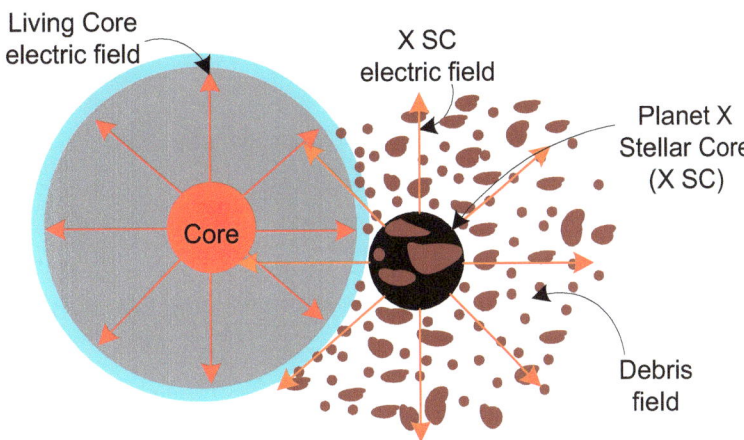

Figure 6.2. A Planet X Stellar Core closely approaches a living celestial object with a living core. Both objects generate a strong positive electric field when the X SC field overlaps, the living core's field, the living core is immersed in an extremely strong electric field, which triggers a matter creation event. Matter explodes from the living core. This matter is in the form of hot liquid magma, which triggers a volcanic eruption (or earthquake), in the case of a planet, and a CME, in the case of a star.

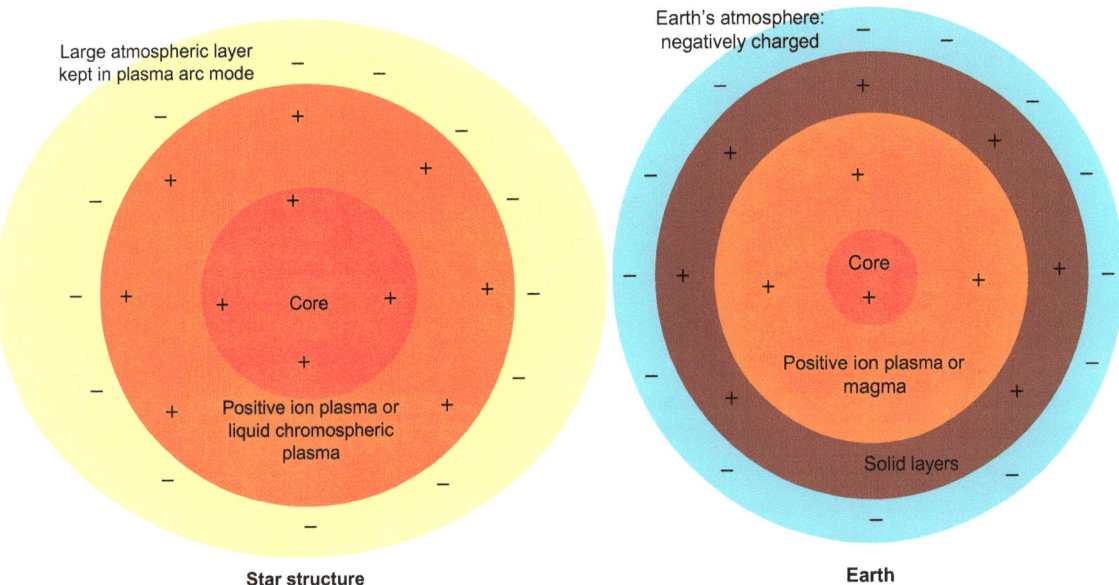

Figure 6.3. The liquid plasma layers go all the way to the surface in stars, but planets have a solidified outer layer. Both have neutral surfaces, positively charged interiors, and negatively charged atmospheres. The negative potential is so high in stars that the atmosphere goes into plasma arc mode and causes the star's light emission. Only very large cores are able to produce such a large outer electric potential

Now, the Planet X System has been affecting the Solar System, at least since the Great Flood, which seems to have occurred between 4000 and 5000 years ago, and is now affecting the Solar System, in an unprecedented way, with the number of objects having reached the Sun in the last, perhaps 50 years, being unprecedented, as large numbers of these objects have been observed in the Sun's corona and the amount of debris dust entering the earth's atmosphere has also reached unprecedented levels. In addition, the numbers of earthquakes and volcanic eruptions have greatly increased, and these can only be happening, if a Planet X Stellar Core of a size close to or larger than earth's core is entering the earth's atmosphere and provoking the earth's core into matter creation events. This seems to be what occurred at the time of the Great Flood, which also caused the earth's surface to have become littered with volcanoes (see Article 515: Current cataclysm engulfing the Earth started at the Great Flood) [5]. The effects however continued over the next thousands of years as the earth continued to have volcanic eruptions, such as the Krakatoa volcanic eruption, in 1883, indicating that these objects, continued to arrive and approach the earth.

Mars seems to have been affected as well, as it too has volcanoes on its surface, a sign that its core was provoked into matter creation event by at least one Planet X Object. Now, we would think with the earth being approached over thousands of years by Planet X System Stellar Cores that its moon would have been affected as well, in which case, there should be volcanoes on the moon. However, there are no volcanoes on the moon, whatsoever. There is evidence of lava flowing on its surface, which is what has given rise to the features called Mares, but there are no volcanoes associated with these features. This, therefore, indicates that the lava, which flowed across the surface of the moon, did not come from the moon's interior and that therefore the moon does not have a living core.

Where would the moon's lava have come from, then? Well in Article 501: Planet X induced volcanic eruptions are like an Earth CME [4], lava can be seen issuing from the Popo volcano, in Mexico, and flowing across a surface above it.

Figure 6.4. Lava seems to be exploding from the volcano, but at the same time lava seems to be exploding from a point above the volcano. This shows that lava attached and flowed across the surface of a Planet X System Stellar Core positioned right above the volcano.

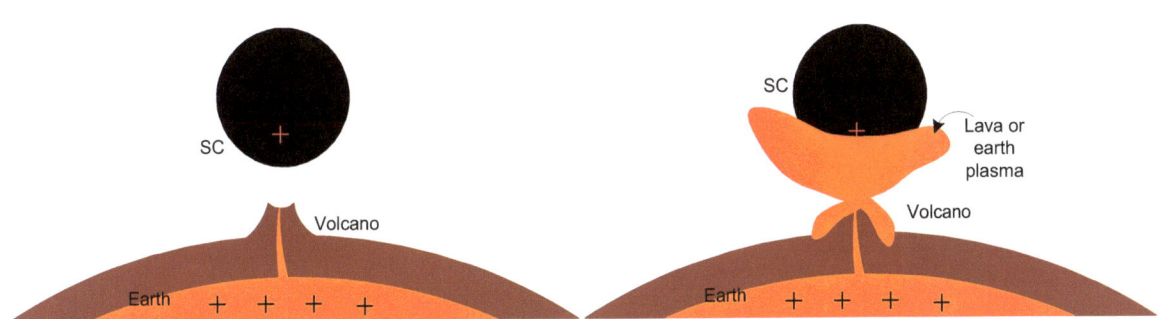

Figure 6.5. A volcanic eruption, like a CME on the Sun, is induced by a large Stellar Core which has a comparable electrical potential to a magma layer inside the earth and induces that layer into having a matter creation event.

This means that the lava flows on the moon came from volcanic eruptions on the earth and the moon must have been inside the earth's atmosphere, at the time, and that therefore the moon was once a Planet X System Stellar Core. Once it had gained a layer of material, on its surface, it was no longer able to closely approach the earth, and must have then settled in the orbit it is in now. It is likely that the layer of material above the dead core is thin, which means that the moon, which is one-sixth the size of the earth, is almost entirely core and that this core is, therefore, a little larger than earth's core, which is one-fifth of the size of the earth. This core would have low gravitational influence as it is dead but a strong electric field which would be able to provoke the earth into stronger matter creation events if such a core where to come as close as to enter the earth's atmosphere as these objects clearly do. The moon could, therefore, be the Planet X System Stellar Core that caused the Great Flood and settled into orbit around the earth soon after the event.

Many would argue that God created the moon when he recreated the earth, as that is what it states in Genesis. However, the evidence, in this case, does not agree with the statement in Genesis which seems to imply that the moon was placed in orbit around the earth then, so it is possible that the translators of the Bible were so convinced that the moon had always been there that they translated the Bible as they did, without it being what was intended in the first place. There is also historical evidence of humans being on earth at a time when there was no moon in the sky. Immanuel Velikovsky found evidence in Greek writings of recorded time in history when there was no moon in the sky. The Greek philosopher, Aristotle, for example, mentions that Arcadia in Greece, at a time before it was inhabited by the Hellenes, was inhabited by a people called Proselenes, when there was no moon in the sky. Plutarch, another Greek philosopher, wrote about the Arcadians as being pre-Lunar people [5].

In conclusion, the moon appears to have a dead core and is, therefore, a Planet X System Stellar Core, which must have once entered the earth's atmosphere, provoked the earth into matter creation events, absorbed lava issuing out of earth volcanoes and eventually settled in a regular orbit around the earth. This seems to have occurred during a time when human beings already inhabited the earth. This and the fact that the moon as a core would be a little larger than earth's core makes it likely that it was the actual Planet X Object, which caused the Great Flood.

References:

[1] Albers, C. (2018). Article 523: Planet X and the Solar System: Jupiter and all gas giants are a recent acquisition.
[2] Albers, C. (2018). Article 514: Stellar Cores are gravitational poles or super proton and white holes.
[3] Albers, C. (2018). Book 3: Planet X Reveals Gravity and Light.
[4] Albers, C. (2018). Article 501: Planet X induced volcanic eruptions are like an Earth CME.
[5] Velikovsky, I. (the 1940s). The earth without the Moon: https://www.varchive.org/itb/sansmoon.htm

Chapter 7

529. Planet X debris field and water clouds

Figure 1 shows a Planet X System Stellar Core surrounded in what appears to be an enormous debris field. The objects seem to all be surrounded in a cloud-like material. Figure 2 shows a huge Planet X System Stellar Core, which turned out to be about 4 times larger than the Sun, which is also surrounded by a huge debris field. We see that the whole of the Sun's corona seems to be filled with the debris and that the pieces of debris also seem to be surrounded in a cloud-like material.

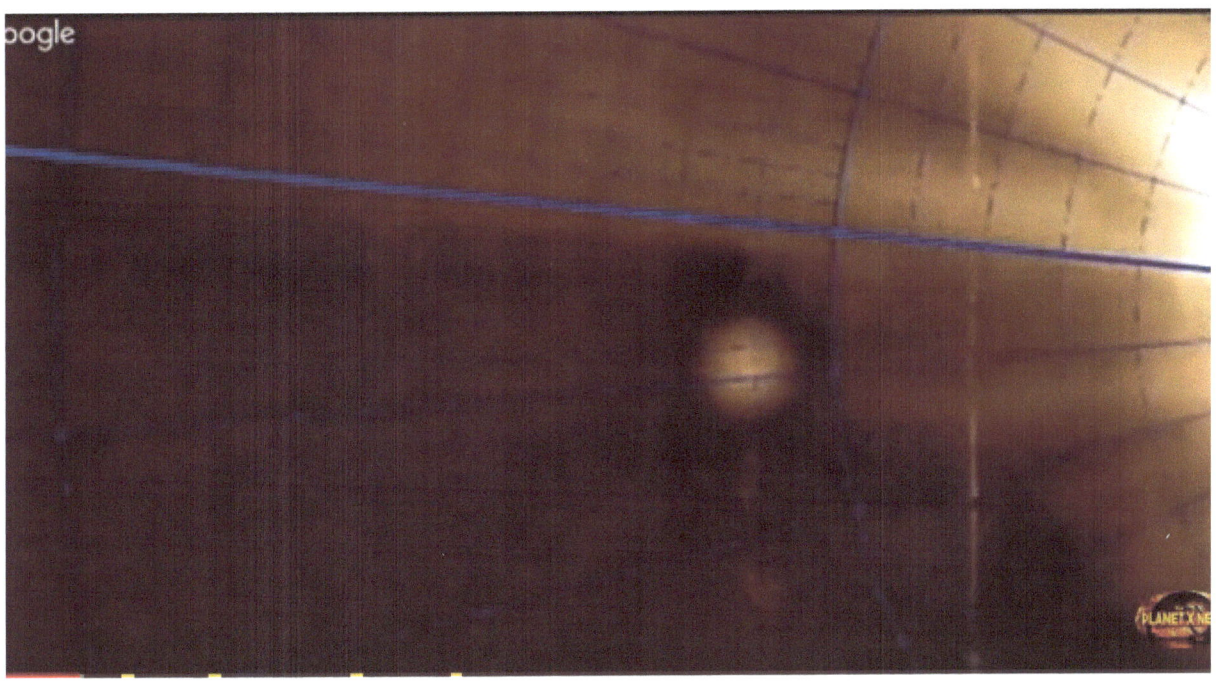

Figure 7.1. This Stereo Hi1 A image of the object shows that it is surrounded and followed by a huge cloud of debris. The debris pieces seem to be surrounded in cloud-like material (see Article 424: Large Planet X Object and large debris field may endanger Earth) [1].

Now, Planet X System objects are depleted in gravitational energy, which gives them low gravitational influence (see Article 210: Stellar Core gravity: tidal and G is not constant) [2], which is why such large objects can enter the Solar System, with no apparent disruption of the planets' orbits around the Sun. They have also been observed traversing the Sun, inside the Sun's corona, at a speed much lower than the Sun's escape velocity, which clearly shows that gravity is not what we have been taught. The strength of the gravitational attraction is dependent on the gravitational photon energy or light within the particles, making up a celestial object (see Book 3: Planet X Reveals Gravity and Light) [3]. In addition, the objects behave like protons interacting through the strong interaction, in the nucleus, because the gravitational and the strong interactions are one and the same interaction (see Article 514: Stellar Cores are gravitational poles or super proton or white holes) [4].

Figure 7.2. Huge Stellar Core within a CME in a Stereo COR2 image from September 13th, 2017 at 7:11 (UTC). The object was at least 4 times larger than the Sun (see Article 321: Huge Planet X star in the inner Solar System) [5]. The object appears to be surrounded in cloud-like structures, which in many cases appears to be toroidal so that a dark circle can be seen in the center. The Sun's corona appears to be filled with these objects and material, which is particularly apparent to the right of the object, and below the Sun.

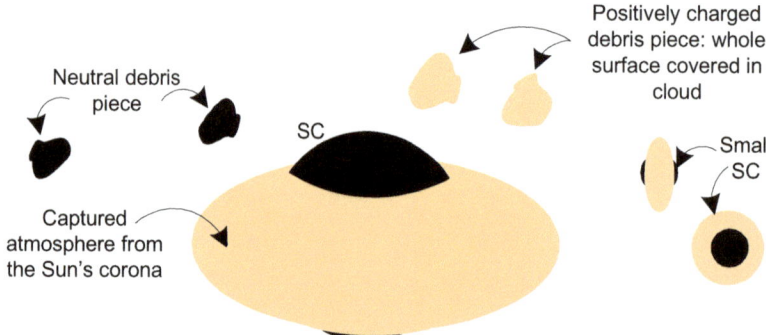

Figure 7.3. The clouds seem to have a toroidal shape as illustrated above. This cloud material has to be water which split into tiny droplets when its gravitational energy dropped. Most of the debris pieces will be positively charged as they come from the interior of the planet and will thus attract debris water in the form of a cloud, as water is a polarized molecule. Therefore these objects and most of the debris pieces will arrive in the Solar System enveloped in cloud.

The reason for my thinking that now is that I now understand that the objects, which are large enough, will be gathering positive ion plasma from the Sun or another Solar System object; it is only the smaller ones that gather atmosphere. This is due to the fact that they seem to only be able to attract material which has slightly less gravitational potential than they do (see Article 576: Planet X larger than the Sun on a collision course with Earth: what happens?) [6]. So, it is likely that the Planet X cores divest themselves of the cloudy material they are surrounded in once they manage to attract a more compatible material from a living celestial object within the Solar System but they arrive in the Solar System with this covering. The debris pieces, which are positively charged would also be covered in it.

Figure 7.4. Stereo A COR2 image showing the Sun's outer corona, which is also full of objects, which are covered in cloud-like material. No large Stellar Core can be seen but the outer corona is full of this material indicating that the Planet X System Stellar Cores leave their debris behind and it remains suspended in the Sun's outer corona, as they enter the Sun's inner corona, and absorb energy from the Sun, causing a huge amount of debris to accumulate around the Sun.

It is likely that space around the earth and the earth's atmosphere is also filled in the same type of material, since the objects are clearly coming to the earth in larger and larger numbers, as evidenced by the increasing number of earthquakes and volcanic eruptions (see Article 501: Planet X induced volcanic eruptions are like an Earth CME) [7].

But what could the cloud like material be? It is obviously low density. We know that clouds found on earth seem to be made of water, could these be clouds made of water? Water is supposed to be solid at zero pressure and temperatures close to 0 K, which is the temperature it would be at, in space. However, this water comes from energy depleted planets and stars and so it will be low in gravitational potential, so it is possible that as a result, the gravitational force between atoms is so much lower than normal that the water does not freeze into the solid phase, in space conditions, but that instead, it forms a cloud. A cloud is actually made of tiny water droplets, it is not gaseous. It is thus likely that these clouds formed when the Planet X System planets broke apart, as a result of their energy depleted state, which then caused liquid water, on the planet, to lose cohesion (gravitational attraction) due to its new low energy state, which then caused the water to break up into tiny droplets. Then since the water molecule is polarized, the negative side attached to the positively charged debris, through the electric interaction, thus forming clouds around the Stellar Cores and the debris pieces.

Since clouds are seen surrounding these objects in space and are seen inside the earth's atmosphere, we can only conclude that the water clouds in our atmosphere are coming in from space and contain water from the Planet X System. This then explains why water, which is much denser than air, can remain suspended in the form of a cloud, from hours to days, in the earth's atmosphere. The clouds are made of gravitational energy deficient water, which takes time to absorb energy in order to get to a potential that will allow it to reach the surface of the earth.

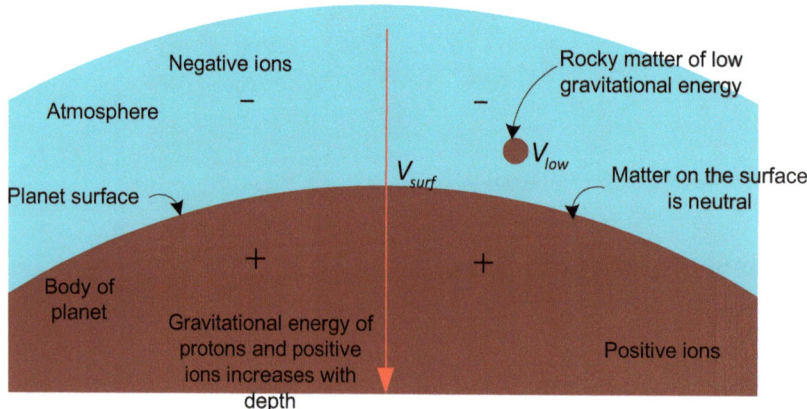

Figure 7.5. Gravitational energy increases with depth and thus decreases with altitude. Matter with less gravitational energy than typical solid matter, at the surface, will remain suspended in the atmosphere. Thus, Planet X debris in the form of dust, pieces of rock or water, which will be low in gravitational energy, will have a lower gravitational potential than matter belonging on the earth's surface, and thus on entering the earth's atmosphere, will remain suspended in the atmosphere, until it has gained enough gravitational energy to reach the surface.

It is true that the 'powers that be' have found a way of producing artificial clouds through chemtrails but it seems that the clouds that have been seen in our atmosphere for thousands of years are coming from outside our atmosphere, and they would also have started coming in at the time of the Great Flood cataclysm (see Article 515: Current cataclysm engulfing the Earth started at the Great Flood) [8].

The fact that rainbows are observed in the earth's atmosphere, from the surface, can only be possible if there are water droplets suspended in the earth's atmosphere, which can only be Planet X water.

The process through which Planet X water absorbs energy inside the earth's atmosphere must involve the absorption of electrons from the earth's atmosphere, and most likely an exchange of electrons, the water gains free electrons from the earth's atmosphere, and the low gravitational energy electrons leave the water, and become free electrons. These free electrons will then absorb light in order to equalize their potential to the environment, at their particular altitude. The absorbed electrons will most likely emit light, as they release excess energy, in order to settle in a certain energy level, which would cause the water cloud to seem to become luminescent. Noctilucent clouds are high altitude luminescent clouds, which seem to have first been noticed in the upper atmosphere in 1850 and are thus the result of Planet X water, in cloud form, entering the earth's atmosphere (see Article 272: Noctilucent clouds and Planet X debris in the earth's atmosphere) [9].

In conclusion, water clouds on earth, which have been observed in our atmosphere for thousands of years, seem to come from space, and to be made of low gravitational energy water and thus, a part of the Planet X System debris fields.

References:

[1]   Albers, C. (2018). Article 424: Large Planet X Object and large debris field may endanger Earth.

[2]   Albers, C. (2018). Article 210: Stellar Core gravity: tidal and G is not consta.

[3]   Albers, C. and C'one, S. (2018). Book 3: Planet X Reveals Gravity and Light.

[4]   Albers, C. (2018). Article 514: Stellar Cores are gravitational poles or super proton or white holes.

[5]   Albers, C. (2018). Article 321: Huge Planet X star in the inner Solar System.

[6]   Albers, C. (2018). Article 576: Planet X larger than the Sun on a collision course with Earth: what happens?

[7]   Albers, C. (2018). Article 501: Planet X induced volcanic eruptions are like an Earth CME.

[8]   Albers, C. (2018). Article 515: Current cataclysm engulfing the Earth started at the Great Flood.

[9]   Albers, C. (2018). Article 272: Noctilucent clouds and Planet X debris in the earth's atmosphere.

Chapter 8

531. Planet X entering earth's atmosphere: creating funnel-shaped clouds

Figure 8.1 below shows a long funnel-shaped cloud, coming from a large cloud and reaching down to the ground. The cloud looks somewhat like what we would expect from a tornado cloud formation, but it has the width of a waterspout formation. Clouds are made of tiny droplets of water, which has low gravitational potential, and which has formed around the Planet X System Stellar Cores and debris pieces. The clouds enter the atmosphere when the objects enter the atmosphere accompanied by their debris fields (see Article 529: Planet X debris field and water clouds) [1]. The Stellar Cores (SC) which come to earth, from this system, are the cores of dead planets; the debris will contain pieces of rock from the planets. The clouds formed from the planets' water and is thus a part of the planets' debris field, which was left over when the planet died or exploded (see Article 513: Planet X planets are exploded planets) [2].

Figure 8.1. Funnel-shaped cloud formation from a MrMBB33 video entitled: 'People looked in AWE as it towered over the city – Rare' from December 31st, 2018 [3].

Water in the form of clouds remains suspended in the atmosphere because of its low gravitational potential, and therefore, when part of the cloud moves down toward the surface, it means that the water, in the cloud, has gained enough gravitational potential to be pulled towards the ground, by the earth's gravitational attraction. Thus, the funnel-shaped cloud is actually a gravitational vortex, and it forms for the same reason that water moving down the through a hole, in a sink, forms a vortex. The effect is due to gravitational diffraction; the amount of water is large compared with the size of the hole and thus the hole acts as a source of the gravitational field. In the case of a cloud vortex, the Planet X System Stellar Core is the source of the gravitational field, even if it is a very weak source, due to its very low gravitational energy state.

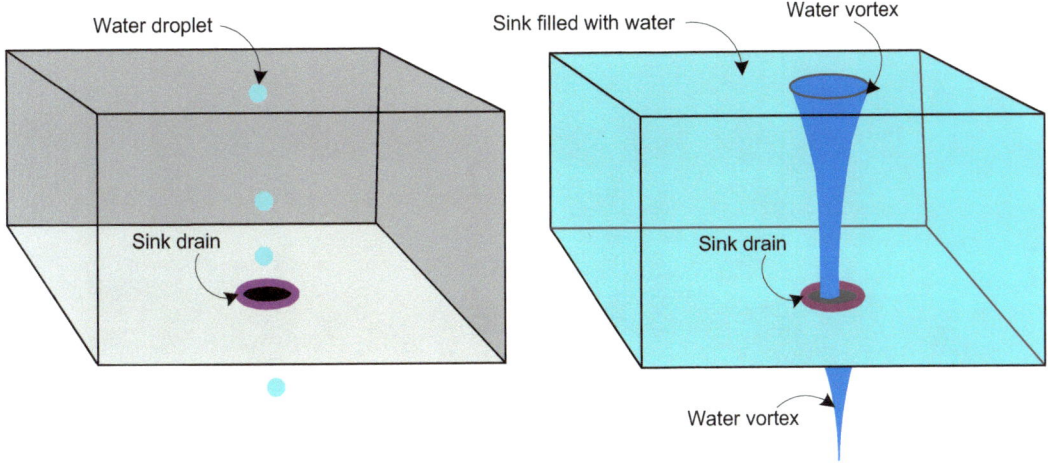

Figure 8.2. A water droplet will fall right through a sink drain, without any gravitational vortex (diffraction) effects appearing. But if the sink is filled with water, the vortex appears and a vortex also forms under the drain. The vortex is widest where the force is strongest and in the case of the sink, it is strongest at the surface of the water. The water going through the hole continues to fall as a vortex after it passes the hole (see Article 380: Gravitational diffraction: gravity is a wave) [4].

Figure 8.3. Funnel-shaped cloud, in Brazil, from a video on the Planet X News Facebook page. The funnel-shaped cloud was rotating.

The funnel-shaped cloud formation above is similar to what we can see in figure 1, but it is much wider. It also did not touch the ground, suggesting that none of the water making up the cloud had enough gravitational potential to reach the surface, but that the water at the bottom of the vortex had the highest gravitational potential and it was getting close to getting the same potential as earth's surface atmospheric material has. The width of the vortex suggests that the object inside the cloud, in figure 2, is larger than the one in figure 1, as a larger vortex would be associated with a larger hole or a larger source of the gravitational field. Figure 3 below shows an even larger funnel-shaped cloud, also from a MrMBB333 video.

Figure 8.4. Large funnel-shaped cloud, i.e. a water vortex in the earth's atmosphere due to the presence of a Planet X System Stellar Core (SC) in the earth's atmosphere.

The above vortex, or funnel-shaped cloud, is even larger suggesting that the Planet X Object, which will be spherical, at the top of the vortex, is even larger. The cloud does not reach as close to the ground, as the bottom of the vortex, in figure 2, so the water, at the bottom of this cloud, has lower gravitational potential than the water at the bottom of the cloud in figure 2.

Figure 8.5. Damage to buildings due to the high winds generated by the cloud gravitational vortex, first seen in figure 1: The low width vortices seem to rotate much faster than the wider ones and thus produce high winds which can damage buildings.

There will also be a low-pressure system associated with each of these objects in the atmosphere due to their weak gravitational field and thus attractive force on the earth's atmosphere around the object.

The process by which the water, in these clouds, gains gravitational energy is through the exchange of electrons with the earth's atmosphere. Atmospheric electrons are high in gravitational potential, which then transfers gravitational potential to the cloud around the object. The gravitational potential can then transfer to the object itself. As the object gains gravitational potential, it may eventually attract material from the earth's surface such as water, thus forming water spouts. Unless the object ever gains enough gravitational potential to form its own outer negative layer, it is not likely to leave the earth's atmosphere.

The water in the clouds around these objects are not likely to however reach the ground or fall as rain as the object gains gravitational potential through the water cloud its gravitational influence will increase and will cause the attractive force on the water cloud toward the Planet X Stellar Core to increase which will cause the funnel to move back up toward the surface of the object. It is only the water surrounding the debris pieces and debris dust which will also be in the form of clouds that will fall to the ground as rain once the water clouds have absorbed enough gravitational energy through electron exchange with the earth's atmosphere. The cloud is darkest where the water is denser and thus where the water droplets are likely to have reached a larger size. This will happen as the cohesive forces between water molecules increases due to increasing gravitational energy in the water molecule particles. When the clouds appear to be pink or colored or luminescent this appears to be due to electrons giving off excess light as they settle in a Planet X water molecule. Clouds which are SC cloud envelopes will eventually break away from the object when it has gained sufficient gravitational energy and electrons and will remain as the clouds we have been seeing for thousands of years and which eventually gain enough gravitational energy and fall as rain.

In conclusion, funnel-shaped clouds in the earth's atmosphere are due to the presence of Planet X System Stellar Cores inside the earth's atmosphere. These water clouds, made of small water droplets of low gravitational energy, surround the objects and the material in their debris fields and since the water was a part of these planets, the clouds are also a part of the Planet X System Stellar Cores' debris fields.

References:

[1]   Albers, C. (2018). Article 529: Planet X debris field and water clouds.
[2]   Albers, C. (2018). Article 513: Planet X planets are exploded planets.
[3]   MrMBB333 video: People looked in AWE as it towered over the city – Rare
      https://www.youtube.com/watch?v=1jL5H-pHIUs
[4]   Albers, C. (2018). Article 380: Gravitational diffraction: gravity is a wave.

Chapter 9

532. Planet X or comet reenergizing process

Planet X System Stellar Cores have been coming into the Solar System, for thousands of years, like comets, and have therefore been affecting the earth and the Sun, for a very long time. These objects go through a process inside the Solar System, which allows them to absorb energy and matter from living celestial objects, in the Solar System, i.e. objects that have plenty of gravitational photon energy in their particles, and thus a living core, capable of both generating a strong gravitational field as well as create matter, whilst the Planet X Stellar Cores are severely depleted in gravitational energy, and are thus not able to create matter (see Article 514: Stellar Cores are gravitational poles or super proton or white holes) [1]. This energy absorbing process allows them to turn into Solar System objects, the larger ones that go to the Sun to absorb energy turn into gas giant planets, and the smaller ones, which have gone to the earth, to absorb energy, turn into earth moons (see Article 523: Planet X and the Solar System: Jupiter and all gas giants are recent acquisitions and Article 526: Planet X and the Moon: the Moon has not always been in sky) [2, 3].

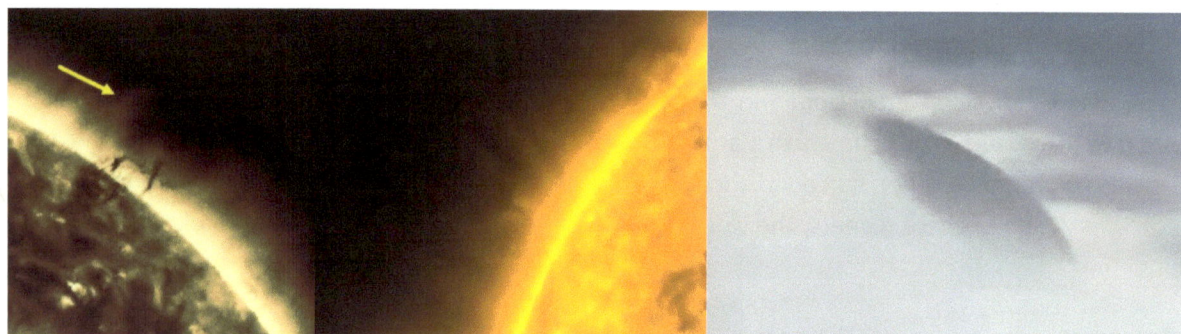

Figure 9.1: Left and center: SDO images of Planet X System Stellar Cores in the Sun's corona: forming gas giant planets. **Right:** Photographic evidence of Stellar Cores within the earth's atmosphere: A spherical object immersed in the cloud: a new moon in the process of forming (Source: R. Wayne Steiger).

Living Stellar Cores generate a high positive electric field, and the Planet X System Stellar Cores retain this strong electric field. It is this electric field, which stops them from colliding with the surface of any other celestial object (see Article 514: Stellar Cores are gravitational poles or super proton) [1]. Thus, the cores of celestial objects are like protons, within the nucleus, and living celestial objects, are like atoms, i.e. superatoms, as they have an outer negative layer. Planet X System Stellar Cores come into the Solar System without an outer negative layer, and they do not have the gravitational energy to produce one, but once they absorb gravitational energy, they will be able to eventually produce a negative outer layer and will thus become whole objects, superatoms. However, their cores will remain dead; they will not have the energy to create matter, which is why there are no volcanoes on the moon, it has a dead core [3].

The cores of celestial objects create matter and therefore all the outer layers of a celestial object are created by the core at the center of that celestial object. Matter emerges in the form of hot positive ion plasma or magma, and as more is created, the older material becomes the outer layers of the object, which eventually cools and solidifies, into a rocky surface, if it is a planet, and becomes the outer chromosphere, if it is a star (see Article 524: Mercury: created by the Sun due to Planet X) [4].

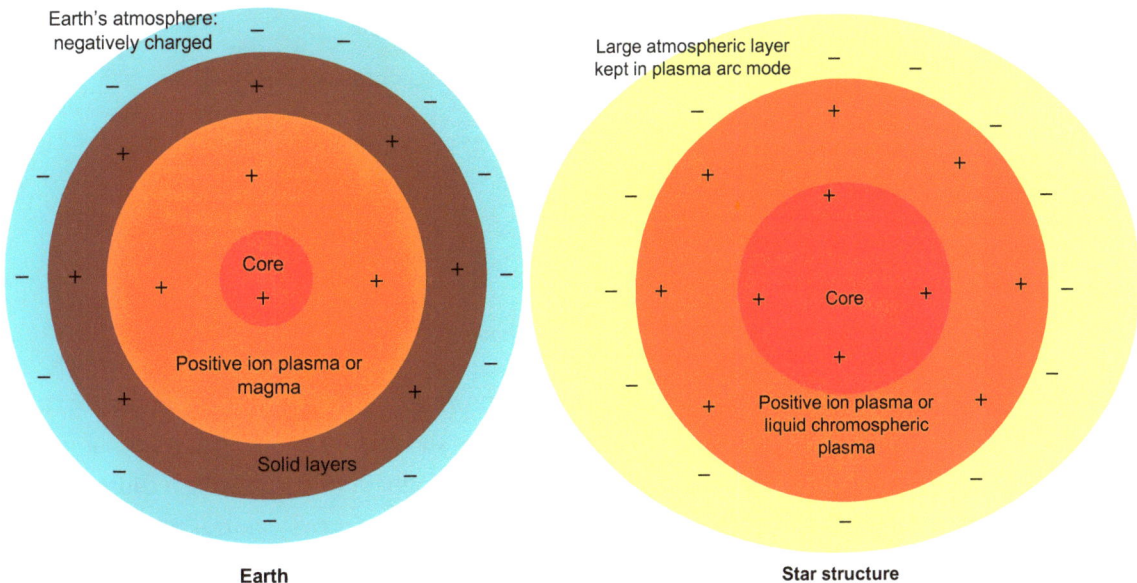

Figure 9.2. The main difference between stars and planets is size; both have a core, are positively charged on the inside and negatively charged on the outside. They both have a liquid layer around the core but this layer goes all the way to surface with a star, whilst a planet has a solid surface and solid rock layers under the surface. Both have gaseous atmospheres but stars generate such a high electric field, in their outer layers, that their atmospheres go into plasma arc mode and thus emit huge amounts of light (see Article 506: Planet X, Earth CMEs and star internal structure) [5].

Lighter elements, within the magma, float upwards, toward the surface, and end up separating from it, so that water ends up, on the surface, and gaseous elements end up as atmosphere. At the same time, the gravitational interaction repels electrons outwards to the outer layers, thus creating a negatively charged outer layer, i.e. the atmosphere will be negatively charged. Since it is the gravitational interaction, the charge separation part of it, which causes protons and electrons to repel, and that leads to the formation of this outer negative layer, and since Planet X System Stellar Cores are low in gravitational potential, they are not able to generate an outer negative layer and thus operate as positive superiors, instead of neutral superatoms, like living celestial objects do.

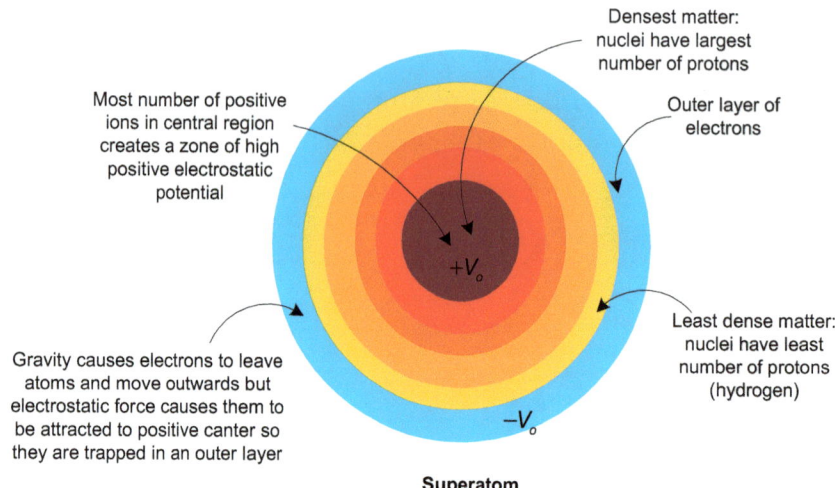

Figure 9.3. The layers making up a celestial object all come from the core, which creates matter, the density of the layers decreases, as we move toward the surface, and the gravitational potential decreases as well. It is the charge separation part of the gravitational interaction, which causes the outer least dense layer, or atmosphere, to be negatively charged; the surface is neutral and the interior is positively charged.

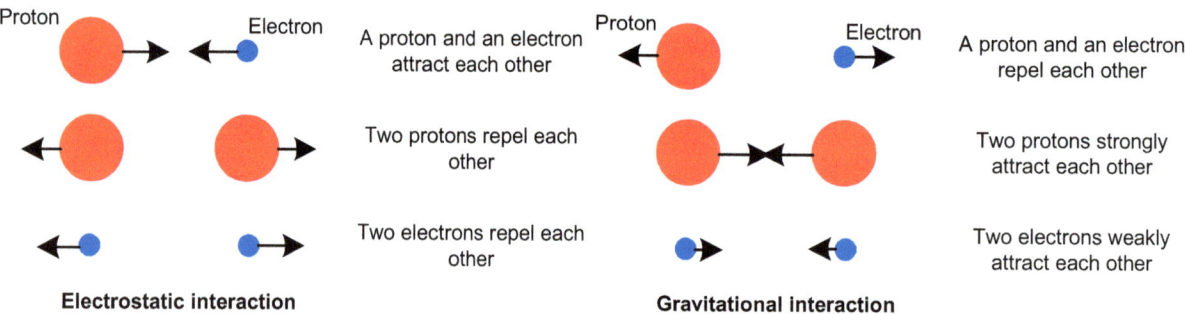

Figure 9.4. The electrostatic and the gravitational interactions between protons and electrons (see Book 3: Planet X Revealed Gravity and Light) [6].

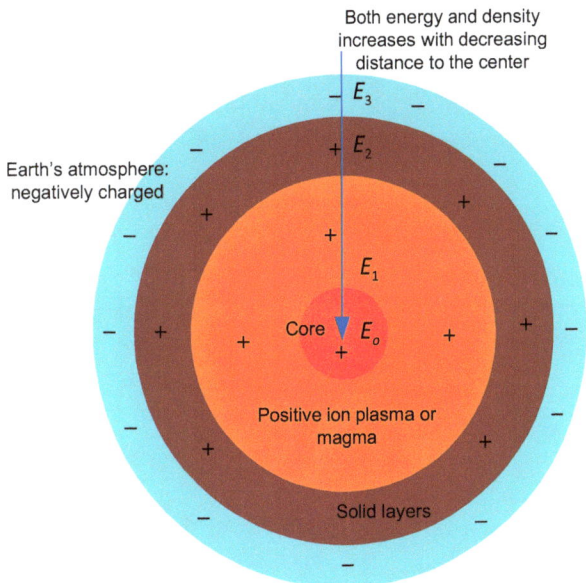

Figure 9.5. Gravitational energy per unit mass, or gravitational potential, increases with depth and is in the form of energy levels, matter, in the core, has the highest gravitational potential, matter in the molten rock region is in the next highest energy level, matter in the solid rock layer is lower in gravitational potential, and matter in the atmosphere is in the last and lowest energy level (see Article 514: Stellar Cores are gravitational poles or superprotons or white holes) [1].

Since their gravitational energy is so low, Planet X System Stellar Cores are only capable of extremely weak gravitational interactions, so it is the electric or electrostatic interaction that brings them to Solar System objects; they are attracted to their negative outer layers. And since the Sun has a huge atmosphere, the heliosphere, which encompasses the whole Solar System, out to 122 au, they would have been attracted to the Sun's outermost layer and would absorb as much energy as they could from that layer. But even though the electrons in the outermost layers would be very high in gravitational energy, the electron cloud at that distance would be very dispersed, because that layer will have the lowest density, resulting in very low energy, per unit mass, and also, per unit volume.

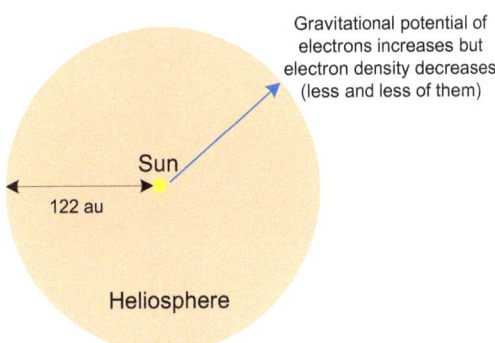

Figure 9.6. The Heliosphere is a spherical bubble, or very low-density atmosphere, around the Sun, which is fed by the Solar Wind. The Solar Wind is made up of matter created by the Sun's core, it will have high energy electrons in it, but fewer and fewer, as the distance from the Sun increases.

Thus, the objects would move inwards toward the Sun, where the gravitational energy per unit volume increases to very high levels. However, there is an effect called gravitational tuning, which does not allow the objects to absorb material with higher gravitational potential than the core, so only the largest Planet X Objects would go to the Sun, where they would find material of compatible gravitational potential, some will only be able to absorb material from the inner corona, or atmosphere, others from different depths of the chromosphere; so, most likely, the ones that had been stars, as living celestial objects, go to the Sun, and the smaller ones, which had been planets, would go to the planets like earth (see Article 576: Planet X larger than the Sun on collision course with Earth: what happens? [7].

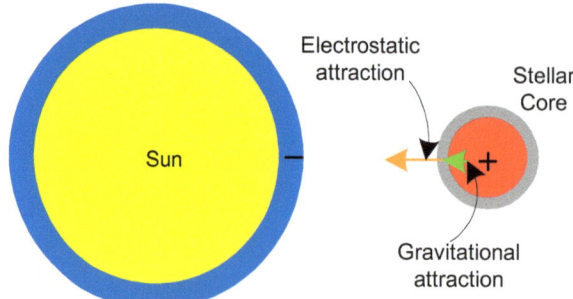

Figure 9.7. It is the electric or electrostatic interaction that brings Stellar Cores to living celestial objects. A Stellar Core because it is positively charged is electrostatically attracted to the negative outer layer of a living celestial object like the Sun.

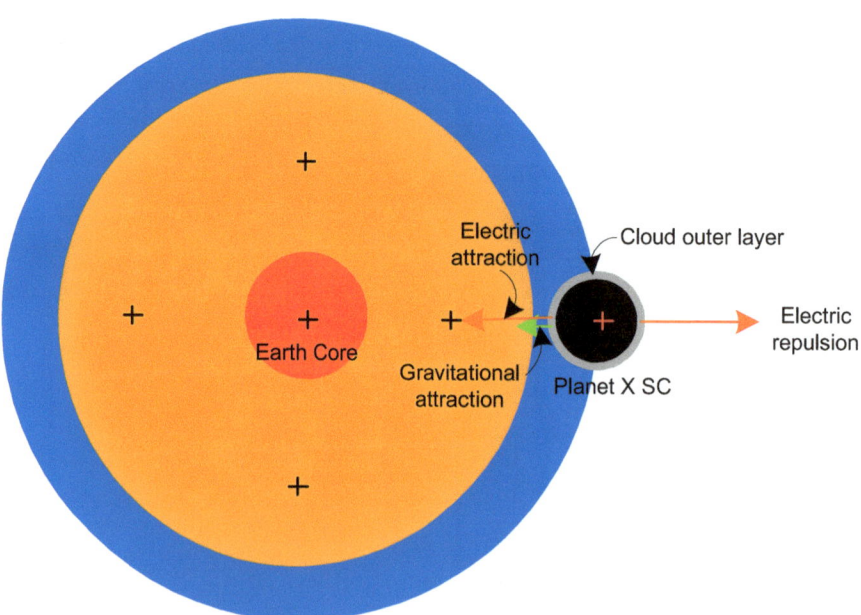

Figure 9.8. Once inside the atmosphere, they will be repelled by the positively charged core inside the living celestial object. However, at that close distance, the gravitational attraction becomes strong enough to be of about the same strength, as the electrostatic repulsion, at some minimum approach distance, which allows the object to hover within the earth's atmosphere or Sun's corona.

Within the atmosphere, the object will absorb material with comparable proton or positive charge density to its core. Some objects are only able to draw electrons in, through their water cloud formations, some are eventually able to draw water from the earth's surface, which is high in gravitational potential and the largest are able to induce the earth to have matter creation events, which gives rise to volcanic eruptions. The magma issuing from the eruption then attaches to the object and the gravitational energy is absorbed by the SC.

Figure 9.9. A volcanic eruption, like a CME on the Sun, is induced by a large Stellar Core which has a comparable electrical potential to a magma layer inside the earth and induces that layer into having a matter creation event.

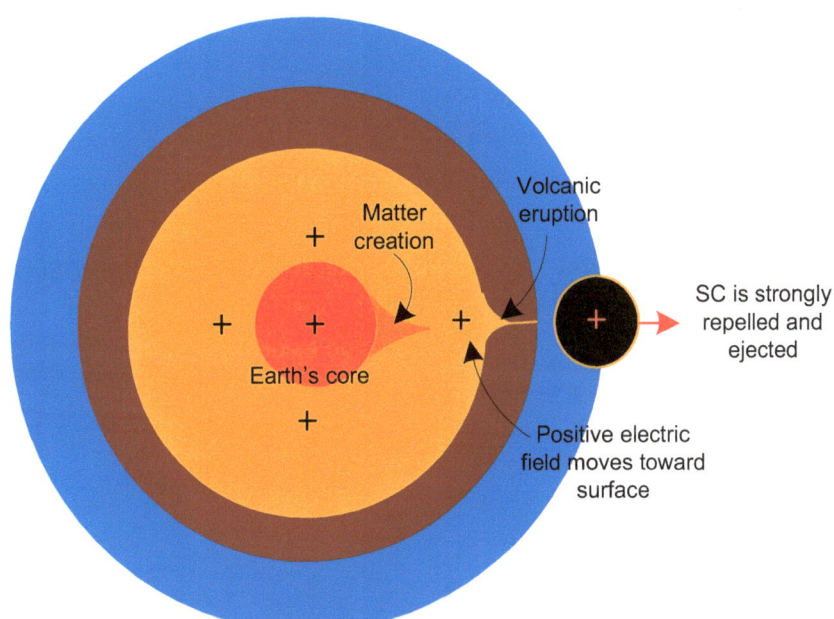

Figure 9.10. The Planet X Object will usually be ejected whenever the host celestial object has a matter creation event because the event distorts the internal positive electric field under the Stellar Core, thus sharply increasing the repulsive electric field.

The larger Stellar Cores will thus often be ejected from the earth's or sun's atmosphere at the time of a volcanic eruption or CME, in the case of the Sun, but it will keep coming back for more material, until they have gained enough gravitational potential to create their own outer negative layer, i.e. have gained enough gravitational potential to repel their own electrons outwards to the outside of their bodies. When this happens their outer negative layer will be repelled by the earth's outer negative layer, and the object will not be able to enter the earth's atmosphere, or, in the case of the Sun, the sun's inner corona, anymore, and will thus go into orbit around its host, at the appropriate orbital distance, for its overall gravitational potential.

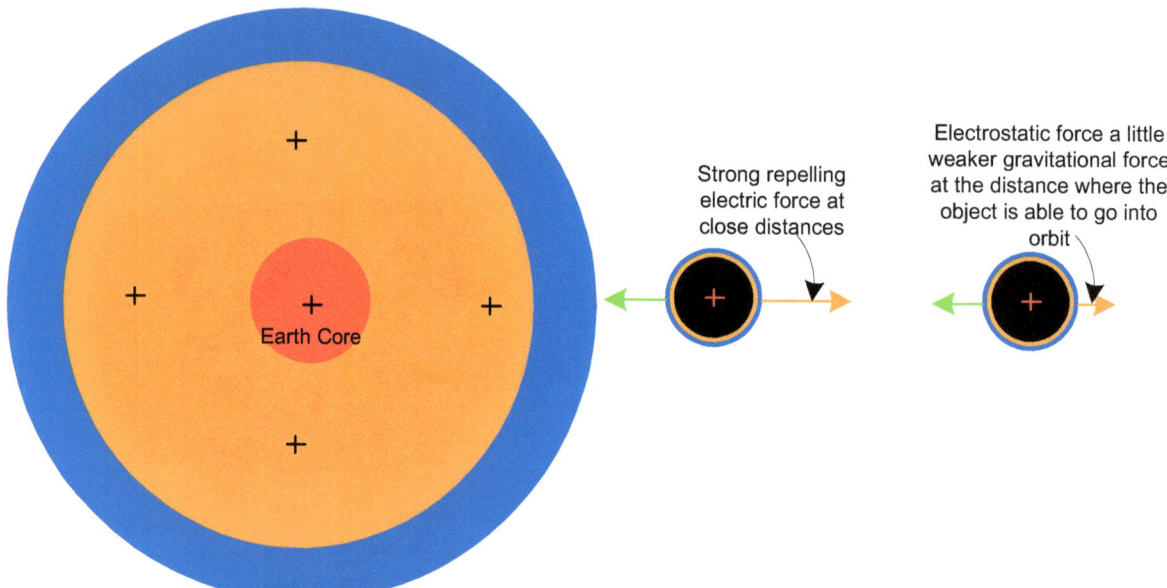

Figure 9.11. With an outer layer of electrons, the Planet X System Stellar Core is now repelled by the earth's outer negative layer instead of being attracted to it and will thus not be able to enter it anymore. It will instead go into orbit at a distance where the gravitational and electric forces are nearly in balance. The electrostatic interaction seems to be extremely strong at very close range (a force which keeps particles from colliding), stronger than the gravitational interaction at very close range. The force which is usually called gravity is actually the sum of the gravitational (strong nuclear force) and the electric force (see Article 514: Stellar Cores are gravitational poles or super proton or white holes) [1].

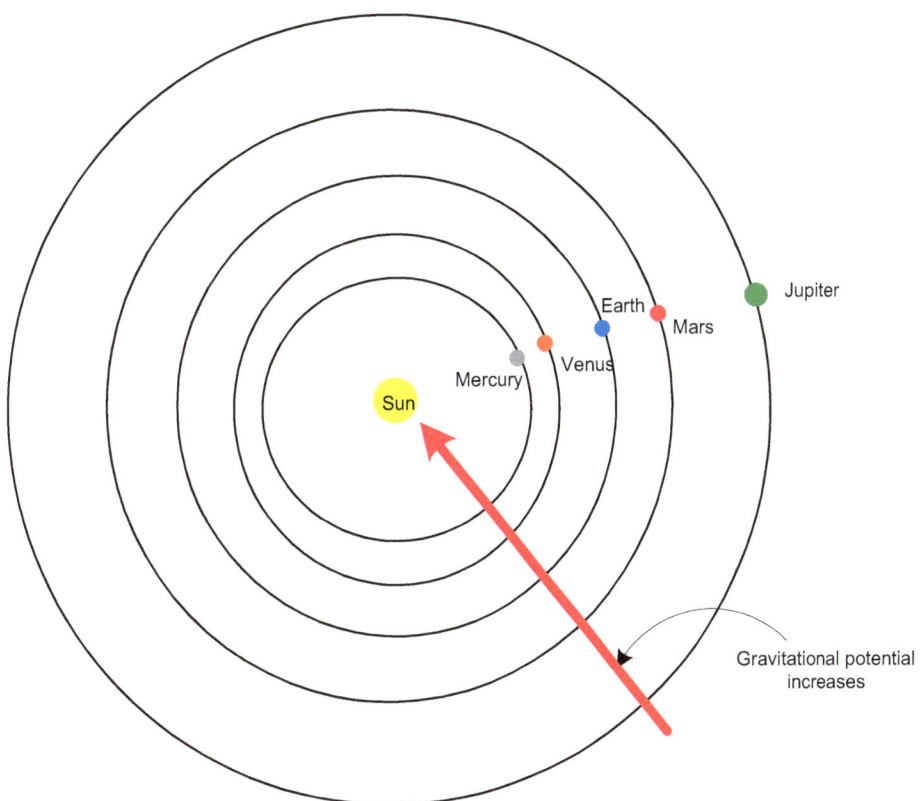

Figure 9.12. Gravitational potential increases with decreasing distance to the sun, so, planets orbiting further out from the Sun have a lower gravitational potential (see Article 523: Planet X and the Solar System: Jupiter and all gas giants are recent acquisitions) [2].

The Planet X Stellar Core will then behave like a normal Solar System object and obey normal gravitational laws but it will seem to be of very low density, as it still has a dead core, which although of higher gravitational potential than before it is still low in comparison with living celestial objects. Thus, even though the object and particularly the core will be large and dense, it will seem to be of very low density and mass, due to its still low gravitational energy status. This is why planets such as Saturn are thought to be made entirely of gas, and why the moon is believed to have a very small core and be of a much lesser density than earth; they are reenergized Planet X System Objects and thus have dead cores. They are also extremely destructive before and during the reenergizing process and thus extremely dangerous to life on earth.

In conclusion, Planet X System Stellar Cores develop into Solar System objects through a process that can be understood in terms of the theory of gravity I have developed based on the observation of these objects in the Sun's Corona.

References:

[1] Albers, C. (2018). Article 514: Stellar Cores are gravitational poles or super proton or white holes.

[2] Albers, C. (2018). Article 523. Planet X and the Solar System: Jupiter and all gas giants are recent acquisitions.

[3] Albers, C. (2018). Article 526: Planet X and the Moon: the Moon has not always been in the sky.

[4] Albers, C. (2018). Article 524: Mercury: created by the Sun due to Planet X.

[5] Albers, C. (2018). Article 506: Planet X, Earth CMEs and star internal structure.

[6] Albers, C. and C'one, S. (2018). Book 3: Planet X Revealed Gravity and Light.

[7] Albers, C. (2018). Article 576: Planet X larger than the Sun on a collision course with Earth: what happens?

Chapter 10

533. Planet X induces super nuclear reactions in the earth's core

Planet X System Stellar Cores are dead planets or stars, which come into the Solar System as comets and end up as gas giant planets, or earth moons. These objects are low in gravitational photon energy and as a result, have low gravitational influence but retain a strong electric field.

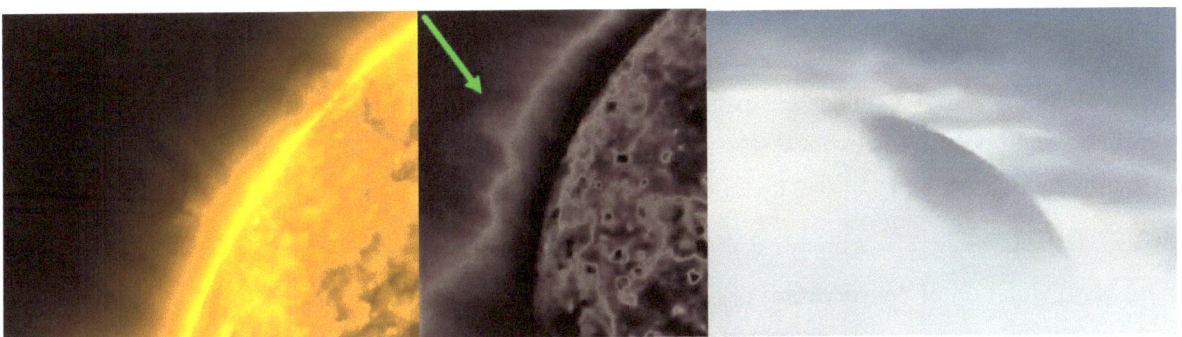

Figure 10.1. Planet X System Stellar Cores in the Sun's corona and in the earth's atmosphere.

These objects are also able to closely approach living celestial bodies, like the Sun and the Earth, to the point that they enter their atmospheres because they are not able to generate an outer negative layer (see Article 532: Planet X or comet reenergizing process) [1]. This allows them to induce a matter creation event in the core of the earth, which results in a volcanic eruption or an earthquake, if the explosion of magma moving outwards toward the surface does not find an outlet, all the way to the surface, in the case of the earth, and which results in a CME event, in the case of the Sun (see Article 501: Planet X induced volcanic eruptions are like an Earth CME) [2].

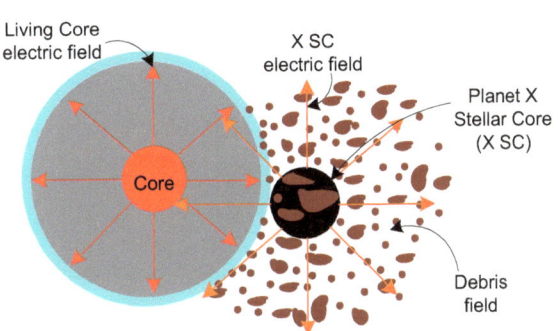

Figure 10.2. A living celestial object's positive field ends at the surface but a Planet X Stellar Core's field does not end, as the object has no negative outer layer, which allows it to enter the atmosphere of a living celestial object. Then as its electric field overlaps the living object's field, the field in the core reaches critical strength and triggers a matter creation event (see Article 526: Planet X and the Moon: the Moon has not always been in the sky) [3].

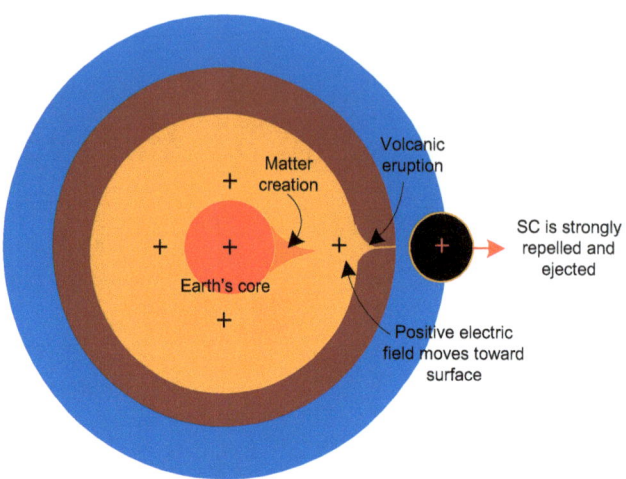

Figure 10.3. The Planet X Object will usually be ejected whenever the host celestial object has a matter creation event because the event distorts the internal positive electric field under the Stellar Core, thus sharply increasing the repulsive electric force on it (see Article 532: Planet X or comet reenergizing process) [1].

There is no violation of conservation of energy; energy is transformed from one form to another only. Light is transformed into matter. Energy can exist in 3 different forms: light or free photons, matter or mass, and gravitational photon energy, which is light when it exists within matter or particles.

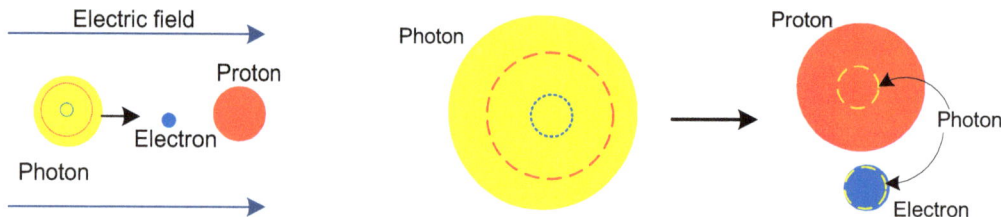

Figure 10.4. Left: A photon, moving through a region of the electric field, split into its constituent particles. The photon energy goes into creating the mass of the particles as well as into gravitational energy, which exists within the particles. **Right:** The dashed lines represent existing particles, within the free photon, with no mass. Solid red and blue circles represent particles with mass. The dashed yellow circles represent photons within particles. The photons within particles are gravitational photon energy on which the gravitational interaction strength is dependent upon (see Book 3: Planet X Revealed Gravity and Light) [4].

Everything comes from within light, so all energy, be it mass, or energy within the mass, is simply light in a different form. Matter, with no energy, or no light within it, will not be able to interact via the gravitational interaction, no matter how much mass it has. In addition, the structure we see at the microscopic (quantum) level, with respect to particles or atoms, is the same that is seen at the astronomical level, with astronomical objects, so that living celestial objects are like atoms, i.e. superatoms, and Planet X System Stellar Cores are like ions (atoms with missing electrons), i.e. superiors. In addition, the core of each celestial object can be thought of as a proton, the nucleus of a hydrogen atom. Living celestial objects have a negative electron layer, and are thus like a neutral hydrogen atom, and a Planet X core is like a hydrogen atom, without an outer electron (see Article 514: Stellar Cores are gravitational poles or super proton or white holes) [5].

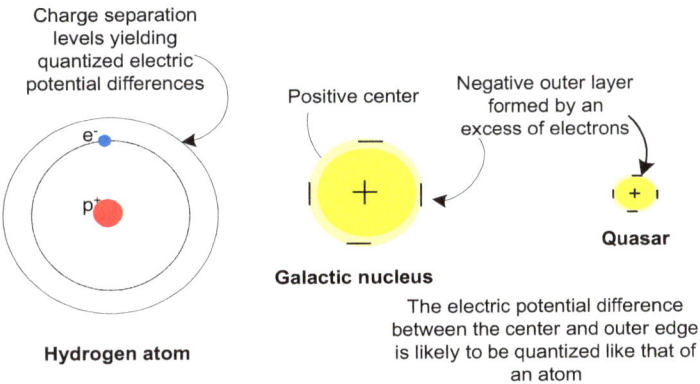

Figure 10.5. Astronomical bodies have the same structure as atoms: dense positive core (nucleus) and a low density outer negatively charged layer (electrons).

Now, if two hydrogen atoms approach each other, they can only get so far, as their outer electron does not allow the two, to come together beyond a certain distance. This is also what happens with celestial objects, once Planet X System Stellar Cores, gain the ability to produce their own outer negative layer, they will never again, enter the atmosphere of another celestial object and will, therefore, go into orbit around their host. But, if a proton approaches a hydrogen atom, it will be able to penetrate it, absorb the electron, which turns it into a neutron and approach the proton, in the nucleus. The result is a nuclear reaction which causes the release of energy in the form of light (free electrons). This is a light creation event, in which some of the light within the protons (gravitational photon energy) is released as free photon energy.

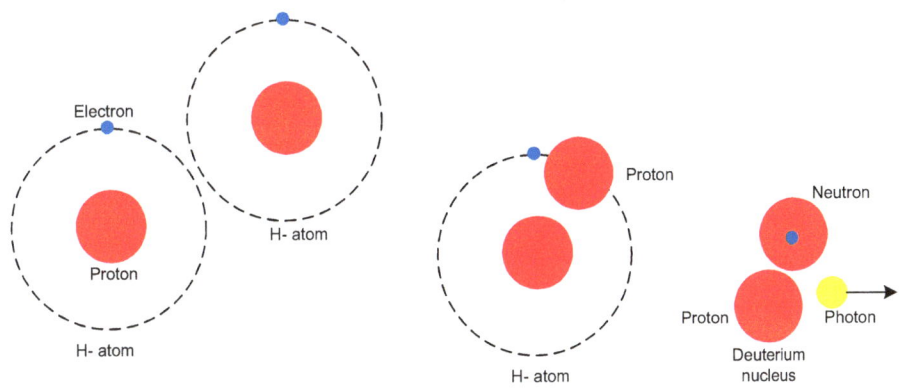

Figure 10.6. Two hydrogen atoms will repel each other, but a proton, on its own, can penetrate the hydrogen atom, which results in a nuclear reaction, in which a light creation event occurs, i.e. energy in the form of a free photon is released.

The reason why gravitational photon energy is released, in such a nuclear reaction, is that the formation of a neutron takes the release of gravitational energy, because a proton and an electron, with high gravitational energy, will repel each other, through the charge separation part of the gravitational interaction; it is only when they have less gravitational energy that the gravitational repulsion decreases to the point that the electric interaction can cause the two (proton and electron) to combine into a neutron.

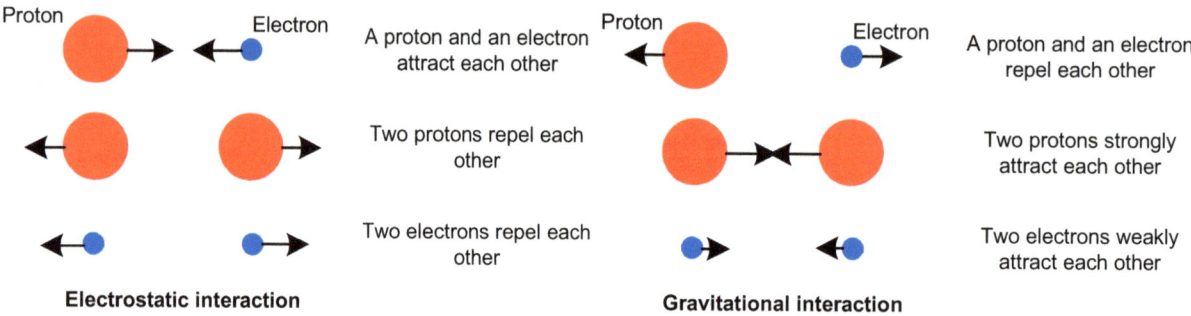

Figure 10.7. The electrostatic and the gravitational interactions between protons and electrons: The charge separation part of the gravitational interaction causes protons and electrons to repel (see Book 3: Planet X Revealed Gravity and Light) [4].

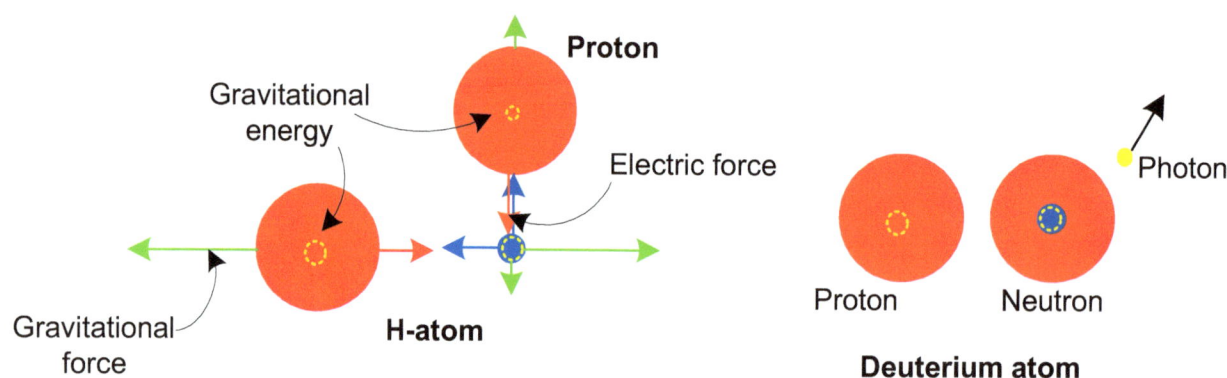

Figure 10.8. When a proton, with less gravitational energy, than an electron, in a hydrogen atom, approaches it, the gravitational repulsion between the two is not strong enough to keep them apart, so they combine into a neutron; some of the photon energy is transferred to mass, for the new particle, and the rest of the excess gravitational energy leaves, in the form of a free photon (the gravitational attraction between the protons is not shown).

In a similar way, when a Planet X System Stellar Core, which has low gravitational energy, approaches the core, or nucleus, of a living celestial object, it triggers the release of gravitational energy, which then gets converted to mass, because of the high electric field environment, as the electric field of both objects overlapping. Thus, a matter creation event induced by a Planet X System Stellar Core is a super nuclear reaction.

The matter creation events, which occur naturally during the formation of a planet, or when a galactic nucleus goes through a matter creation event are also super nuclear reactions (see Article 524: Mercury: created by the Sun due to Planet X) [6]. In this case, the object's gravitational energy is too high for its size, which as a result, also generates a very strong electric field. The object's gravitational energy is too high for the amount of mass available, so the core becomes unstable, which then triggers a release of gravitational energy. In a newly formed planet, for example, this will occur because the newly formed planet will initially have the same gravitational potential as the star that formed it, but the planet's mass is much less than the star's, so gravitational energy is released in the form of free photons, until a state of balance for the size or mass or the planet is reached. Whenever photons emerge as a result of the matter creation events, they emerge into a very high electric field environment, which then causes matter to emerge from within the photons, i.e. matter creation events occur.

The astronomer Halton Arp deduced, through his observations, which he describes in his book 'Seeing Red' [7] that active galactic nuclei were having matter creation events. They were ejecting matter in a fluid form, which condensed into quasars, which then unfurled arms, and became galaxies, in their own right. This is exactly the same process which occurs in stars and planets, all cores are able to create matter (see Article 522: Stellar Cores are sources of matter or white holes) [8]. Galaxies, which were ejecting quasars, were super bright indicating that they had an excess of gravitational energy, for their mass, which then triggered matter creation events.

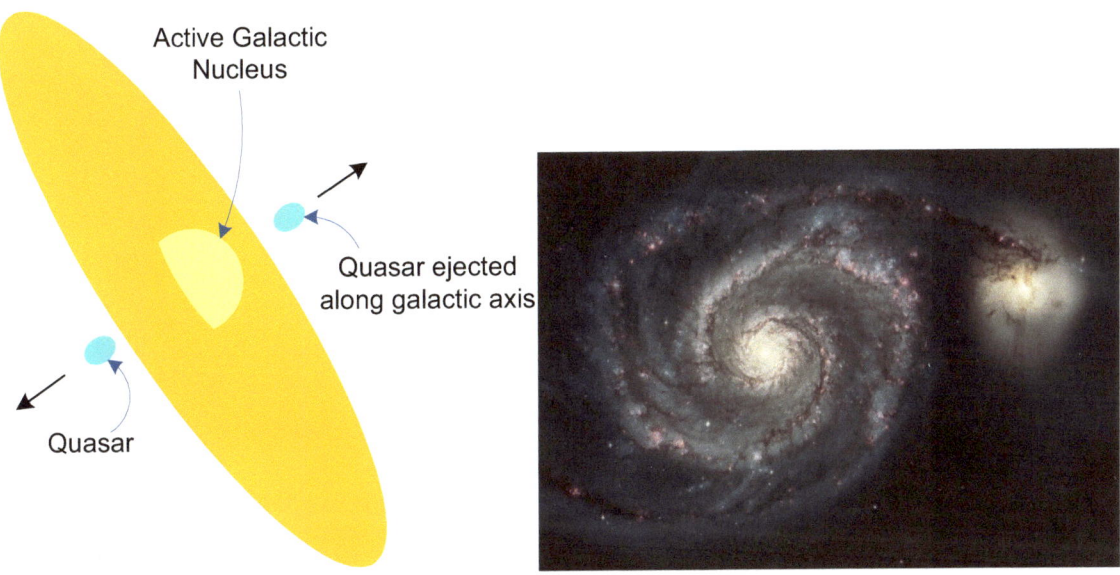

Figure 10.9. On the left: Active Galactic Nuclei are intensely bright galactic centers, with extremely high electric fields. When the electric field and brightness reaches a critical level, it causes instability and the ejection of matter, which condenses into a quasar. In the course of time, the quasars start ejecting their own material, along with their major axis, which develops into arms and thus becomes galaxies themselves. The material ejected from the center of quasars spreads out in a spiral because of the rotational motion of the quasar. Thus galaxies give birth to galaxies, and the matter is continuously being created in the universe. On the right: A spiral galaxy, the spiral arms are due to the material being ejected along the galaxy's plane of rotation (see Article 126: White Holes instead of Black Holes at the Center of Galaxies) [9].

In conclusion, Planet X System Stellar Core observations have led to the understanding that matter creation events, which give rise to volcanic eruptions, on the earth, and CMEs, on the Sun, are like nuclear reactions, which occur at the particle level, they are thus, at the celestial object level, super nuclear reactions.

References:

[1] Albers, C. (2018). Article 532: Planet X or comet reenergizing process.
[2] Albers, C. (2018). Article 501: Planet X induced volcanic eruptions are like an Earth CME.
[3] Albers, C. (2018). Article 526: Planet X and the Moon: the Moon has not always been in the sky.
[4] Albers, C. (2018). Book 3: Planet X Revealed Gravity and Light.
[5] Albers, C. (2018). Article 514: Stellar Cores are gravitational poles or super proton.
[6] Albers, C. (2018). Article 524: Mercury: created by the Sun due to Planet X.
[7] Arp, Halton (1998). *Seeing Red*. Apeiron, Montreal.
[8] Albers, C. (2018). Article 522: Stellar Cores are sources of matter or white holes.
[9] Albers, C. (2018). Article 126: White Holes instead of Black Holes at the Center of Galaxies.

Chapter 11

535. The Sun is no longer shining: what has happened to it?

As I have shown in many previous articles, the Sun is no longer shining (see Article 500: The Sun is no longer shining review) [1]. In this article, I will discuss the reason why the Sun is no longer shining, which appears to be tied to the Planet X System of dead stars and planets, which have been coming into the Solar System as comets for thousands of years (see Article 367: Planet X coming in as comets and affecting the Earth) [2]. These objects are made up of a core surrounded by a debris field.

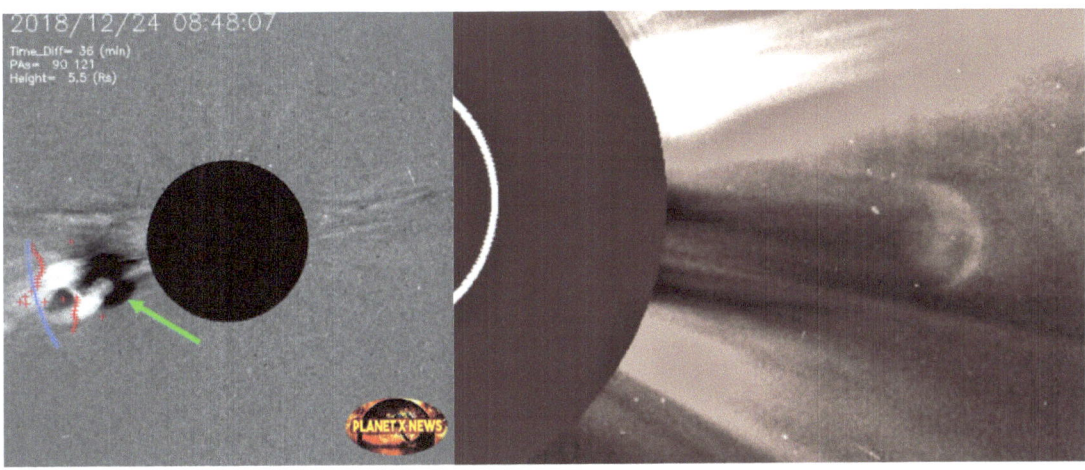

Figure 11.1. Planet X System Stellar Cores in the Sun's outer corona (Source: Scott C'one from Planet X News).

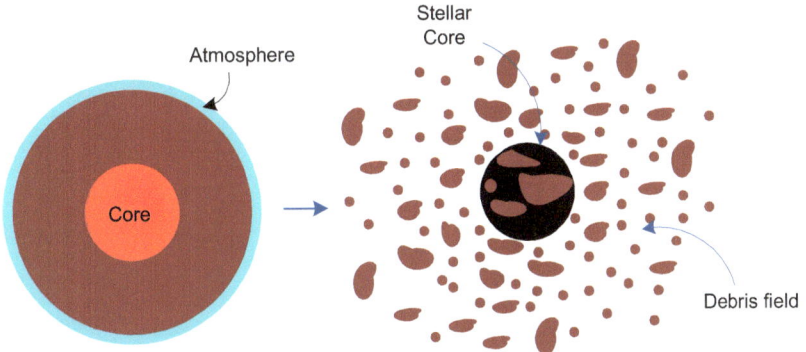

Figure 11.2. A living celestial object turns into a Planet X System Stellar Core, or a dead core, when it becomes low in gravitational photon energy, which seems to have happened suddenly in the case of the Planet X System, as the planets appear to have exploded, when there was still animal life on them (see Article 513: Planet X planets are exploded planets) [3].

The Planet X System Stellar Cores, which go to the Sun seem to go through an energizing process, which turns them into gas giant planets (see Article 523: Planet X and the Solar System: Jupiter and all gas giants are recent acquisitions) [4]. However, they will leave large amounts of this debris, which is low in gravitational energy behind as they go through the process. These debris pieces will remain suspended in the Sun's atmosphere, until they have absorbed enough energy, to reach the Sun's surface; just like the debris entering the earth's atmosphere will also remain suspended in the earth's atmosphere, until it has absorbed enough energy to reach the surface. At the Sun's surface, the debris should simply be absorbed by the Sun's liquid chromosphere.

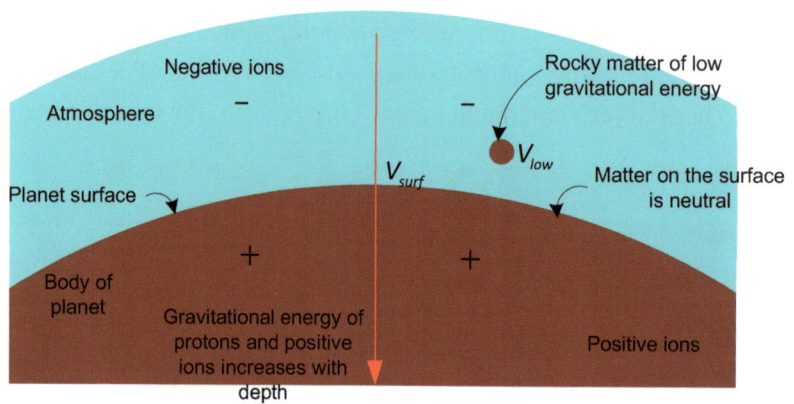

Figure 11.3. Gravitational energy and gravitational potential (gravitational energy per unit mass) increases with depth and thus decreases with altitude. The matter with less gravitational potential than typical solid matter, at the surface, will remain suspended in the atmosphere. Thus, Planet X debris in the form of dust, pieces of rock or water, which will be low in gravitational energy, will have a lower gravitational potential than matter belonging on the earth's surface and will remain suspended in the atmosphere, until it has gained enough gravitational energy to reach the surface.

Since these objects have been coming in for thousands of years, this is most likely what has happened for thousands of years, but either the number of objects has greatly increased recently or the process has slowed due to the Sun's weakness so that the amount of debris seems to have hugely accumulated.

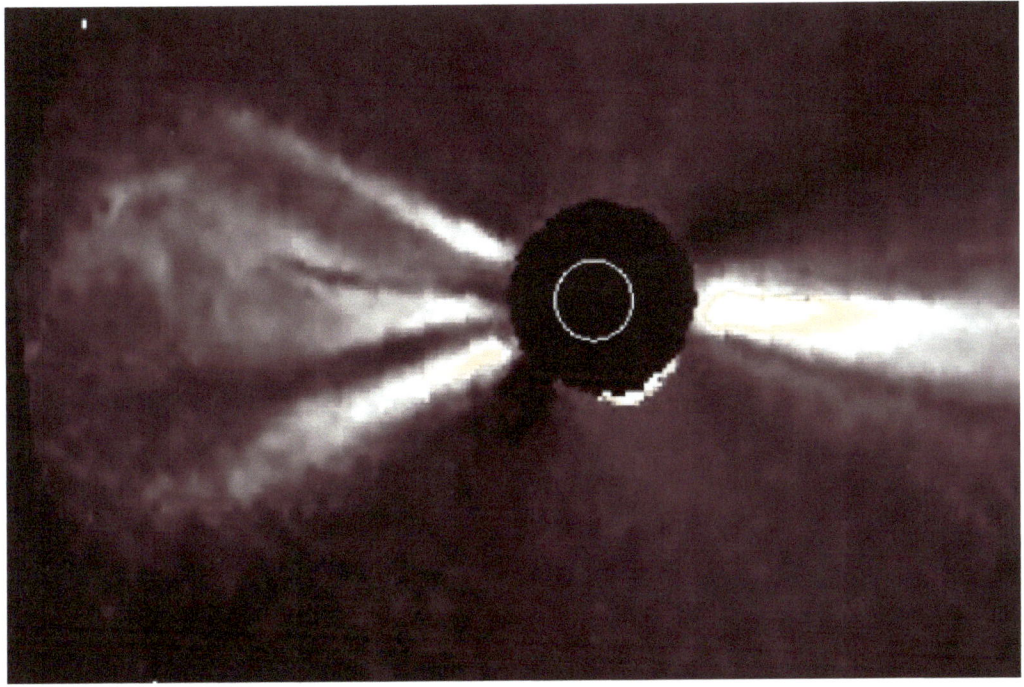

Figure 11.4. Huge Stellar Core within a CME in a Stereo COR2 image, at least 4 times larger than the Sun (see Article 321: Huge Planet X star in the inner Solar System) [5]. The object appears to be surrounded in cloud-like structures, which in many cases appears to be toroidal so that some clouds appear to have a dark center. The Sun's corona appears to be filled with these objects and material. This is particularly apparent to the right of the object, and below the Sun (see Article 529: Planet X debris field and water clouds) [6].

Thus, the Sun's outer corona seems to be full of objects and debris pieces, surrounded in cloud envelopes. These cloud envelopes are a part of the debris and are made of water, which used to be a part of the celestial objects that died. The water is low in gravitational energy, which has caused it to have low cohesion and thus not to freeze solid even in the cold environment of space (see Article 529: Planet X debris field and water clouds) [6].

The debris pieces absorb energy through the cloud envelopes, by exchanging electrons with the Sun's corona and by directly absorbing light emitted by the Sun, just like the clouds entering into the earth's atmosphere do (see Article 534: Planet X in the earth's atmosphere) [7]. And it is most likely the electron exchange that has led to the Sun losing its ability to emit light from its atmosphere, because as electrons with high amounts of gravitational energy, in the Sun's corona, are exchanged for electrons, with low amounts of gravitational photon energy, in the water clouds, more and freer electrons in the sun's atmosphere will no longer be repelled by the core, to the higher altitudes, so that only the lower altitudes would go into plasma arc mode. This would result in the corona seeming to shrink in size and disappear entirely in places, which would look like dark holes in the Sun, i.e. coronal holes. This is what SDO images were showing was happening in the 7-year span of time shown below.

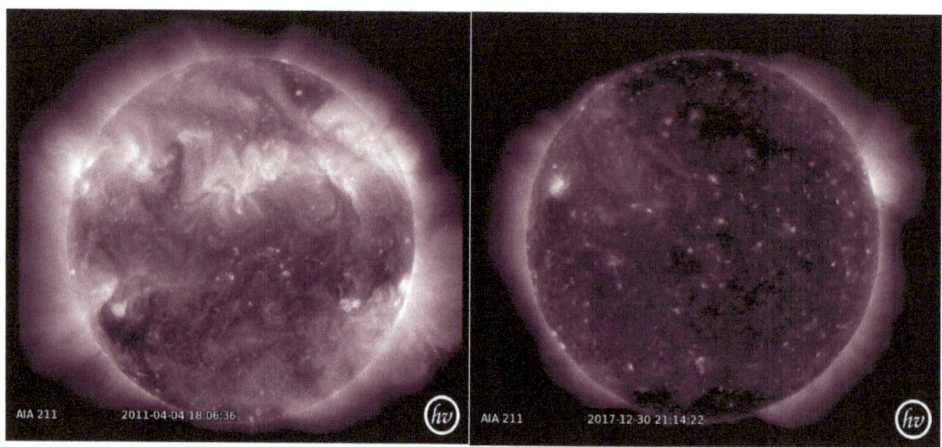

Figure 11.5. SDO images in the 211-angstrom wavelength from April 4th, 2011 and from December 30th, 2017. It is obvious from these images that the Sun lost a lot of corona over the almost 7-year time span. These images are however much older than the time stamps would suggest and so this loss of corona most likely happened a few years before the Sun finally went dark (see Article 500: The Sun is no longer shining review) [1].

Thus, the corona would seem to develop larger and larger coronal holes and eventually, the corona would disappear entirely. But before that would happen, a suddenly increased amount of debris entering the part of the Sun's atmosphere, capable of being in plasma arc mode, would cause the Sun to go dark, in that region. The Sun may have been able to recover from this, at first, so that that the darkness was temporary. A really large influx might even cause the whole Sun to go dark, as it did during the SDO eclipse season, which since it occurred according to a regular period showed that the objects did not remain in the Sun's atmosphere permanently, whilst going through the energizing process, but that they orbited the Sun (see Article 373: Planet X System orbits) [8].

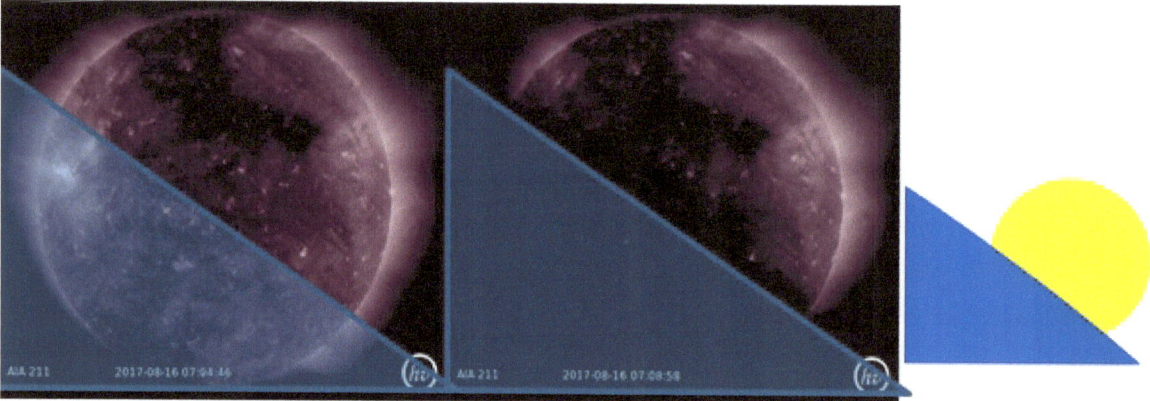

Figure 11.6. The SDO (Solar Dynamics Observatory) eclipse season is supposed to be due to the earth eclipsing SDO's view of the Sun. But the fact that the Sun's corona shrinks back, as darkness progresses across the face of the earth, indicates that this is no eclipse, the Sun is going dark (see Article 324: Planet X causes the sun to be darkened) [9].

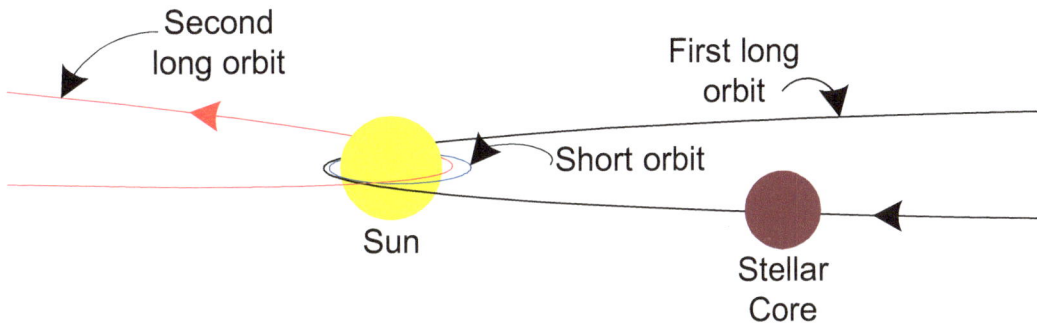

Figure 11.7. SDO eclipse season was a regular cycle and had to be associated to one object arriving at the Sun with its debris field: The shorter orbital period is 177 to 178 days and the longer period is 183 to 184 days, long. The orbit in between is extremely short, with the Sun reaching perihelion once every 24 hours for 24 days, at which time the Sun goes dark for up to 69 minutes [8].

These were temporary but indicated that the Sun's supply of high energy photons was dwindling and that it was only a matter of time before the Sun would go dark for good. Once the Sun was no longer able to place any of its atmospheres, into plasma arc mode, it became a dark star or a planet.

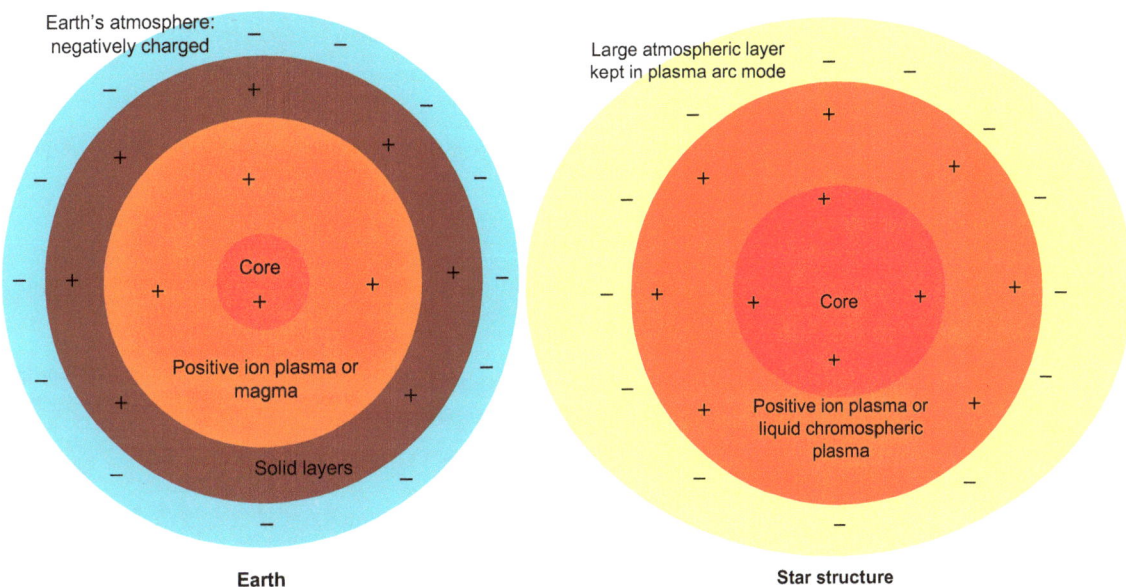

Figure 11.8. The main difference between stars and planets is size; both have a core, are positively charged on the inside and negatively charged on the outside. They both have a liquid layer around the core but this layer goes all the way to the surface with a star, whilst a planet has a solid surface and solid rock layers under the surface. Both have gaseous atmospheres but stars generate such a high electric field, in their outer layers, that their atmospheres go into plasma arc mode and thus emit huge amounts of light (see Article 501: Planet X induced volcanic eruptions are like an Earth CME) [10].

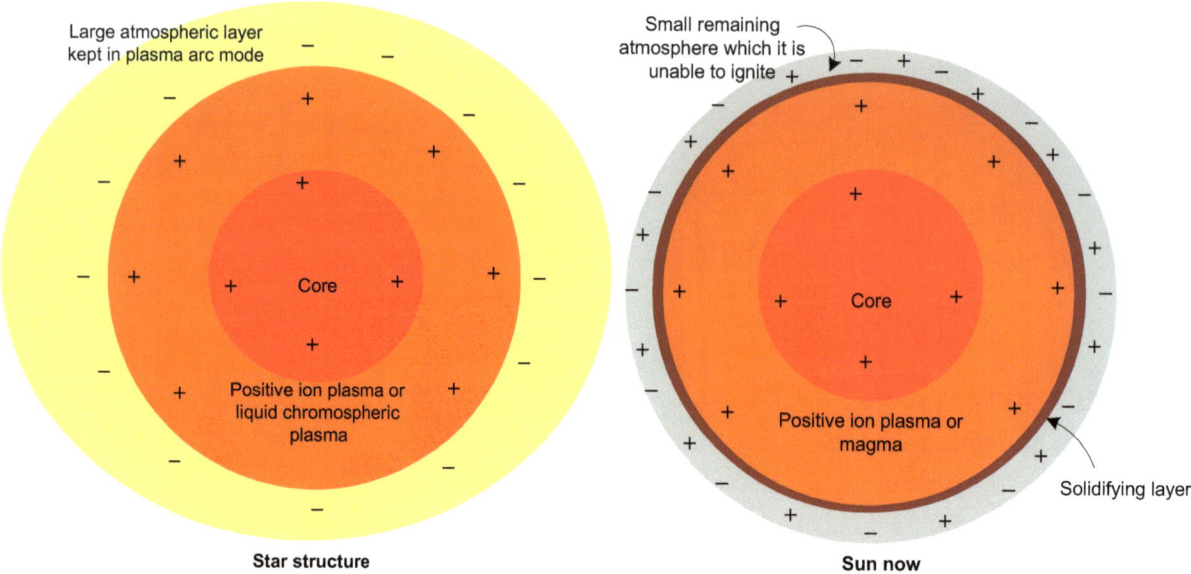

Figure 11.9. The Sun's atmosphere having lost its highest energy electrons, and thus highest altitude electrons, will lose the strong outer electric field, it had before. Thus, the Sun will no longer be able to place its atmosphere into plasma arc mode, the atmosphere will become dark and cold, the surface will thus solidify, and the debris coming in and reaching the surface will accumulate on the solidifying outer layer. The Sun would thus turn into a dark star or a planet [10].

Even as a dark star, the Sun will still be capable of CME events and solar flares, as the Planet X Objects will continue to approach it, and they induce matter creation events in the Sun's core which result in CMEs, but these will now be like earth's volcanic eruptions, just on a larger scale [10].

In conclusion, the Sun is no longer shining, most likely due to the Planet X System of Stellar Cores. These objects have been coming to the Sun for thousands of years, with their debris fields, causing the debris in the Sun's atmosphere to accumulate. This debris absorbs energy through electron exchange, which would cause the Sun's outer electric field to decline, resulting in a decrease in the size of the Sun's atmosphere capable of emitting light, which would be observed as a shrinking corona.

References:

[1]   Albers, C. (2018). Article 500: The Sun is no longer shining review.
[2]   Albers, C. (2018). Article 367: Planet X coming in as comets and affecting the Earth.
[3]   Albers, C. (2018). Article 513: Planet X planets are exploded play.
[4]   Albers, C. (2018). Article 523: Planet X and the Solar System: Jupiter and all gas giants are recent acquisitions.
[5]   Albers, C. (2018). Article 321: Huge Planet X star in the inner Solar System.
[6]   Albers, C. (2018). Article 529: Planet X debris field and water clouds.
[7]   Albers, C. (2018). Article 534: Planet X in the earth's atmosphere.
[8]   Albers, C. (2018). Article 373: Planet X System orbits.
[9]   Albers, C. (2018). Article 324: Planet X causes the sun to be darkened.
[10] Albers, C. (2018). Article 501: Planet X induced volcanic eruptions are like an Earth CME.

Chapter 12

534. Planet X in the earth's atmosphere

The number of Planet X System Stellar Cores, which are dead stars, or planets, entering the earth's atmosphere seems to be increasing, as their effects seem to be increasing. The effects are increased clouds and rain, volcanic eruptions, earthquakes, fissures and landslides, tidal surges and ocean recession events, rogue waves and low-pressure weather systems, such as tornadoes, waterspouts, and hurricanes. It should be surprising that celestial objects approaching earth from space are actually entering the earth's atmosphere, and yet the evidence for this is now overwhelming. This should however not be surprising because there is clear evidence of these objects going into the Sun's corona, which is the Sun's atmosphere. See Article 532: Planet X or comet reenergizing process [1] for the full explanation of why these objects approach living celestial objects, like the Sun and the earth, so closely that they enter their atmospheres.

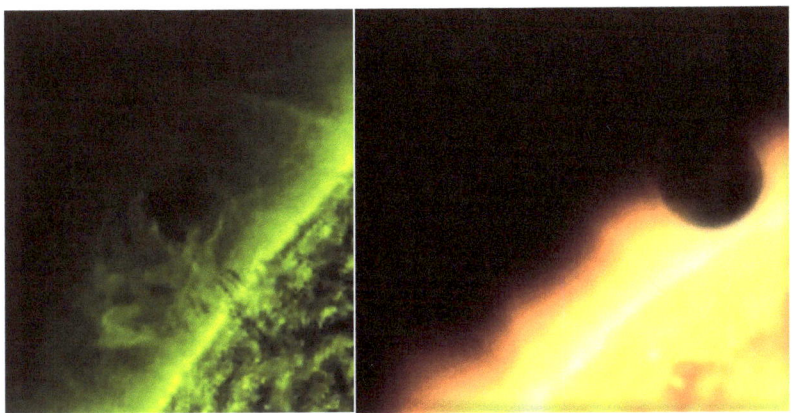

Figure 12.1. Planet X System Stellar Cores in the Sun's corona.

Figure 12.2. Planet X Object immersed in its cloud envelope, within the earth's atmosphere. The cloud formed from the water, which was once a part of the living celestial object that this Stellar Core was a part of before it was destroyed (see Article 529: Planet X debris field and water clouds) [2].

Figure 12.3. The convex shape of what at first appears to be a hole in the clouds indicates that this is a large spherical object, poking through the cloud layer. The cloud layer is pink in places and grey in others indicating that it is absorbing light (see Article 479: Planet X creating circular holes in clouds) [3]. The pink part of the cloud is absorbing green light since pink is a mixture of blue and red and this is what is getting to the camera from this cloud. The grey parts are absorbing all wavelengths.

Figure 12.4. Red, blue and green are the primary colors of light. Red and blue produce pink (see Article 300: Additive and subtractive color theories) [4].

The cloud, around this object, is actually its outer envelope, or outer layer, and is made of small droplets of water. The water has too little gravitational photon energy to be more cohesive and freeze solid, even in the coldness of space.

Figure 12.5. This Hi1 A image of the object shows that it is surrounded and followed by a huge cloud of debris. The debris pieces seem to be surrounded in cloud-like material showing that the cloud envelope is still a cloud even in space (see Article 424: Large Planet X Object and large debris field may endanger Earth) [5].

In the earth's atmosphere, the cloud envelope, around the object, absorbs energy, through electron exchange; the gravitational energy in the electrons then transfers to the core, which thus, increases its gravitational energy. Electrons will sometimes emit some of their excess energy, as they settle into the Planet X water atoms, which can cause the clouds to become luminescent, or even emit different colors of light. The cloud does however also absorb light or free electrons.

Figure 12.6. The spherical shape of the object poking, through the clouds, and giving rise to the water vortex can be clearly seen.

The object's cloud covering, in the above image, is dark blue indicating that the cloud is absorbing red and green light leaving blue to be reflected. This indicates that the cloud absorbs energy through electron exchange as well as through direct light (free photon) absorption. The cloud at the bottom of the vortex is darker because it must be denser at that point, in other words, the water at the bottom, or close to the surface of the planet, has more gravitational energy, than the water further up, in the vortex. When the water making up a cloud has higher gravitational energy, it forms larger drops of water.

In conclusion, Planet X System Stellar Cores, surrounded by a cloud envelope, are entering the earth's atmosphere in large numbers. They absorb energy in the form of light and gravitational photon energy through their cloud envelopes.

References:

[1]   Albers, C. (2018). Article 532: Planet X or comet reenergizing process.
[2]   Albers, C. (2018). Article 529: Planet X debris field and water clouds.
[3]   Albers, C. (2018). Article 479: Planet X creating circular holes in clouds.
[4]   Albers, C. (2018). Article 300: Additive and subtractive color theories.
[5]   Albers, C. (2018). Article 424: Large Planet X Object and large debris field may endanger Earth.

Chapter 13

541. Planet X cover-up and planetary formation: where do asteroids come from?

Planet X System Stellar Core observations have led to the conclusion that these objects are the remains of once-living stars and planets, which have died and are now the cores, hence the name Stellar Core, of these celestial bodies surrounded by a debris field, which used to be the outer layers of these objects. These objects come into the Solar System and become energized by absorbing material from living Solar System objects, like the Sun and the Earth. They induce matter creation events in the cores of these living celestial objects. These matter creation events lead to the emergence of matter in a liquid plasma (magma) form, from the core of the living celestial objects and the explosive emergence of this matter causes CME events, on the Sun, and volcanic eruptions, on the earth (see Article 501: Planet X induced volcanic eruptions are like an Earth CME) [1].

Figure 13.1. This Stereo Hi1 A image of a Planet X System Stellar Core surrounded and followed by a huge cloud of debris. The debris pieces seem to be surrounded in cloud-like material (see Article 424: Large Planet X Object and large debris field may endanger Earth) [2].

The Planet X System objects absorb the liquid plasma and gain gravitational photon energy, by doing so, and eventually manage to produce their own outer negative layer thus turning into reenergized celestial objects (see Article 532: Planet X or comet reenergizing process) [3]. However, they will still have dead cores; the larger ones, which went to the Sun, turn into gas giants, and the smaller ones, which came to the earth turn into earth moons (see Article 523: Planet X and the Solar System: Jupiter and all gas giants are recent acquisitions and Article 526: Planet X and the Moon: the Moon has not always been in sky) [4, 5]. But even the smaller objects, which are not able to induce matter creation events, in the earth's core, will absorb the material in a fluid form; they either absorb atmosphere or liquid water from the earth's surface (waterspouts). In addition, the objects come into the solar system surrounded in water, which is in the form of small droplets of water, and thus in the liquid phase, and therefore in the form, which we call clouds (see figure 1). The water came from the planets that died and is now a part of the debris field, and the densest fluid material that the cores can hold on to (see Article 529: Planet X debris field and water clouds) [6].

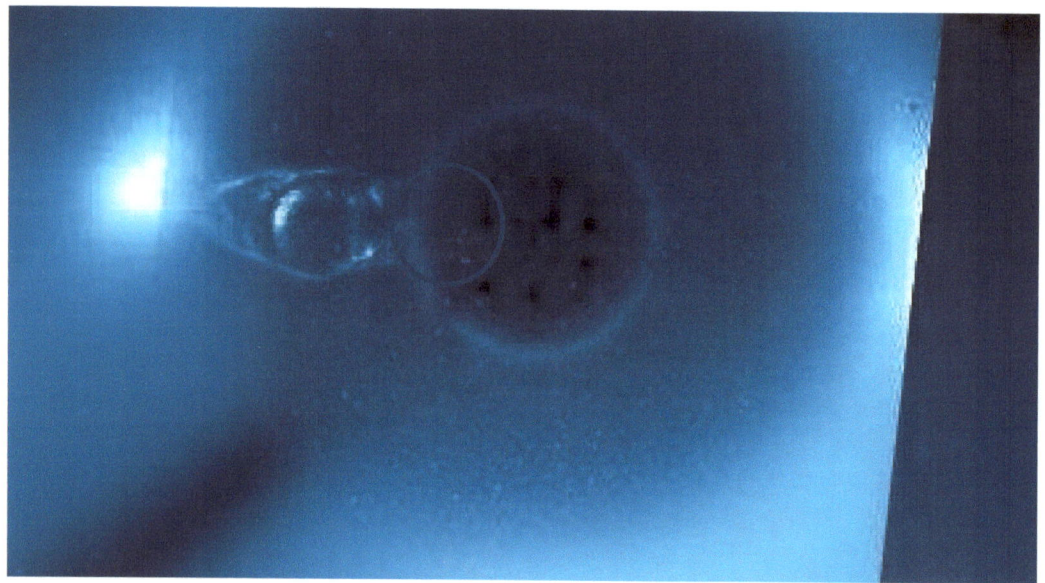

Figure 13.2. Planet X System Stellar Core appears to be about two thirds the size of the Sun ($0.67 r_s$) and is, therefore, a very large core. The CME plasma emerging from the Sun is clearly in the form of a liquid. It is positive ion plasma, which all living cores of celestial objects create (see Article 538: Planet X star appears in a LASCO composite image: liquid CME plasma) [7].

The living cores of celestial objects are able to have CMEs and volcanic eruptions because planets and stars form from the inside outwards. Planets start as fluid ejected by a star in a matter creation event, the liquid plasma condenses into a core, after emerging, from the parent star and then starts creating its own matter, which is also in a liquid form, and which turns into the outer layers of the planet. The material separates according to density with the lowest density material moving toward the surface so that even surface water and atmospheric gases emerge from within. In Article 524: Mercury: created by the Sun due to Planet X [8], I showed that since Mercury is the planet closest to the Sun, it must be the planet with the most gravitational potential, in the Solar System, and that it must, therefore, have a very large core, estimated to be 80% of the whole planet, which then indicates that Mercury is a new planet that recently emerged from within the Sun and is still in the process of forming itself through creating its own matter.

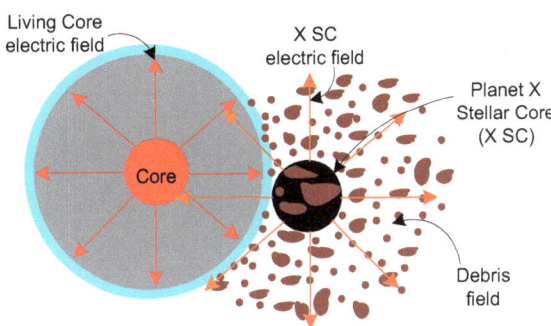

Figure 13.3. A living celestial object's positive field ends at the surface but a Planet X Stellar Core's field does not end, as the object has no negative outer layer, which allows it to enter the atmosphere of a living celestial object. Then, as its electric field overlaps the living object's field, the field in the core reaches critical strength and the core becomes unstable resulting in a matter creation event (see Article 526: Planet X and the Moon: the Moon has not always been in the sky) [5].

Thus, Planet X System observations show that planets emerge from within stars, form from the inside outwards through the cores creating the outer layers of the planet. This conclusion agrees with the astronomer Halton Arp's observations of galaxies and quasars, which showed that quasars were proto-galaxies and that they condensed from material in the liquid phase, which was ejected by active galactic nuclei. These quasars then proceeded to create their own matter, which condensed into star clusters and stars, which become the arms of the galaxy. Halton Arp carefully detailed his observations and conclusions based on these in his book Seeing Red [9]. These have led me to conclude that at the center of galaxies are white holes instead of black holes as matter issues from the center of galaxies and what is occurring is actually the exact opposite of what accepted astrophysics believes is occurring (see Article 126: White Holes instead of Black Holes at the Center of Galaxies) [10]. Now, it turns out that it is not just galactic nuclei that are white holes, all stars and all living planets also are white holes because of they all form and eject their own matter from the inside (see Article 522: Stellar Cores are sources of matter or white holes) [11]. The universe works the same way at all levels. The same mechanism giving rise to nuclear reactions also gives rise to matter creation events (see Article 533: Planet X induces super nuclear reactions in the earth's core) [12].

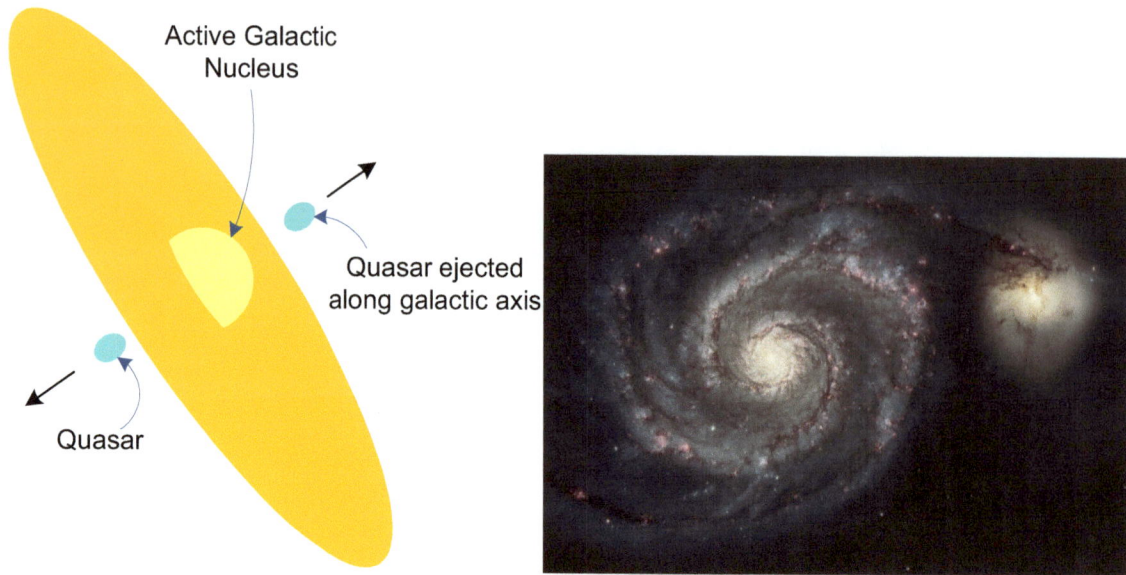

Figure 13.4. On the left: Active Galactic Nuclei are intensely bright galactic centers, with extremely high electric fields. When the electric field and brightness reaches a critical level, it causes instability and the ejection of matter, which condenses into a quasar. In the course of time, the quasars start ejecting their own material, along with their major axis, which develops into arms and thus becomes galaxies themselves. The material ejected from the center of quasars spreads out in a spiral because of the rotational motion of the quasar. Thus galaxies give birth to galaxies, and the matter is continuously being created in the universe. On the right: A spiral galaxy, the spiral arms are due to the material being ejected along the galaxy's plane of rotation.

This shows that the theory I have developed is self-consistent and logical; it does not have contradictions in it; it does not have contradictions because it is based on observation. Now, the accepted theory on planetary formation is that pieces of dust and rock come together in a process called accretion, which eventually gets large enough to be a planet. This theory as I have shown in Article 540. The accretion theory of planetary formation is impossible) [13]. In addition, if accretion was a real process leading to planetary formation, why have not the pieces of rock, and dust, in the asteroid belt, turned into a planet? They have not because those pieces of rock come from the debris fields of the Planet X System objects; they are pieces of broken up planets. But the truth regarding the Planet X System has been covered-up for thousands of years and it continues to be covered-up to this day. It is for this reason that most of the physics research, understanding, and teaching, is wrong and illogical. These illogical theories, full of contradictions, are meant to cause confusion and stop real understanding from emerging.

These incorrect and illogical theories and concepts are meant to keep the understanding of what is actually happening in the Solar System regarding the Planet X System from the earth's population and particularly from the scientific community. This is why any scientist who deviates and starts understanding and getting closer to the truth is ejected from the scientific community or killed. James McCanney, who found, and wrote in his book, Planet X, Comets and Earth Changes [14], that comets turn into planets, was fired from his academic position, in spite of being a wonderful teacher and doing brilliant published research, Dr. Robert Harrington, who most likely saw one of these comets, which are the same as Planet X Objects, coming in, surrounded by their huge debris field, was killed, Dr. Eugene Shoemaker, a planetary scientist and co-discoverer of the comet Shoemaker-Levy 9, which impacted Jupiter, and who must have realized that Jupiter was not just made of gas, but that it must have a solid surface, just below the clouds of gas, was also killed. The reason why Jupiter must have a huge solid body, which must extend to just below the clouds, we see as its surface, is that it is a reenergized Planet X Object, and most likely the core of what was once a small star. Jupiter's core is dead and thus very low in gravitational energy, which makes the object seem to have much less mass and thus have much less density than it actually has (see Article 523: Planet X and the Solar System: Jupiter and all gas giants are recent acquisitions) [4].

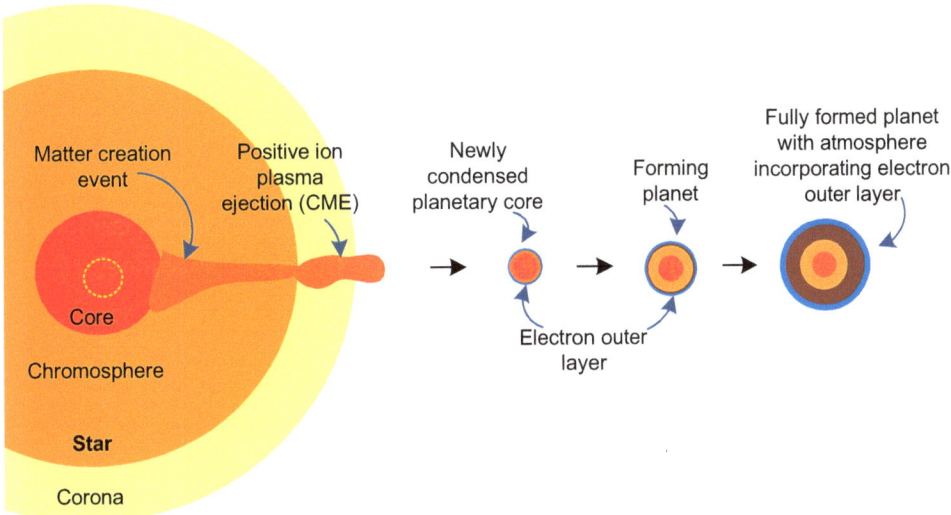

Figure 13.5. Planet X observation based planetary formation which agrees with Halton Arp's galaxy observations. A star ejects liquid plasma which condenses into a core which then ejects its own material so that the outer layers of the planet form.

In conclusion, planets are ejected by stars; they are ejected as hot liquid plasma and condense into a core, which then ejects its own liquid plasma, which then turns into the outer layers of the planet; the same mechanism, which was observed by Halton Arp in galaxy formation. Planetary formation due to accretion is an illogical theory, which seems designed to cover-up the presence and effects of the Planet X System on the Solar System. Asteroids in the Solar System come from the debris fields of the Planet X System; they are pieces of broken up planets.

References:

[1] Albers, C. (2018). Article 501: Planet X induced volcanic eruptions are like an Earth CMEs.
[2] Albers, C. (2018). Article 424: Large Planet X Object and large debris field may endanger Earth.
[3] Albers, C. (2018). Article 532: Planet X or comet reenergizing process.
[4] Albers, C. (2018). Article 523: Planet X and the Solar System: Jupiter and all gas giants are recent acquisitions.
[5] Albers, C. (2018). Article 526: Planet X and the Moon: the Moon has not always been in the sky.
[6] Albers, C. (2018). Article 529: Planet X debris field and water clouds.
[7] Albers, C. (2018). Article 538: Planet X star appears in a LASCO composite image: liquid CME plasma.
[8] Albers, C. (2018). Article 524: Mercury: created by the Sun due to Planet X.
[9] Arp, Halton (1998). *Seeing Red*. Apeiron, Montreal.
[10] Albers, C. (2018). Article 126: White Holes instead of Black Holes at the Center of Galaxies.
[11] Albers, C. (2018). Article 522: Stellar Cores are sources of matter or white holes.
[12] Albers, C. (2018). Article 533: Planet X induces super nuclear reactions in the earth's core.
[13] Albers, C. (2018). Article 540: The accretion theory of planetary formation is impossible.
[14] McCanney, J. (2002). Planet X Comets and Earth Changes. Jmccanneyscience.com press Minneapolis.

Chapter 14

542. Planet X observation based energy conversion in the cores of planets and stars

Planets emerge from within stars (see Article 541: Planet X cover-up and planetary formation: where do asteroids come from?) [1] They are the result of a matter creation event inside a star. Stars and planets create both matter and give off light. They do so by converting gravitational photon energy, which is photon energy, within particles, into both matter and light, in the form of free photons. The matter emerges in the liquid form, from within the core, which suggests that the core itself is not solid; the core is also in the liquid phase, like the magma which it creates but is hotter and denser than the magma that is outside the core. The Planet X System cores, however, seem to be solid, which is to be expected since they are too low on energy to be able to convert light into matter (see Article 532: Planet X or comet reenergizing process) [2].

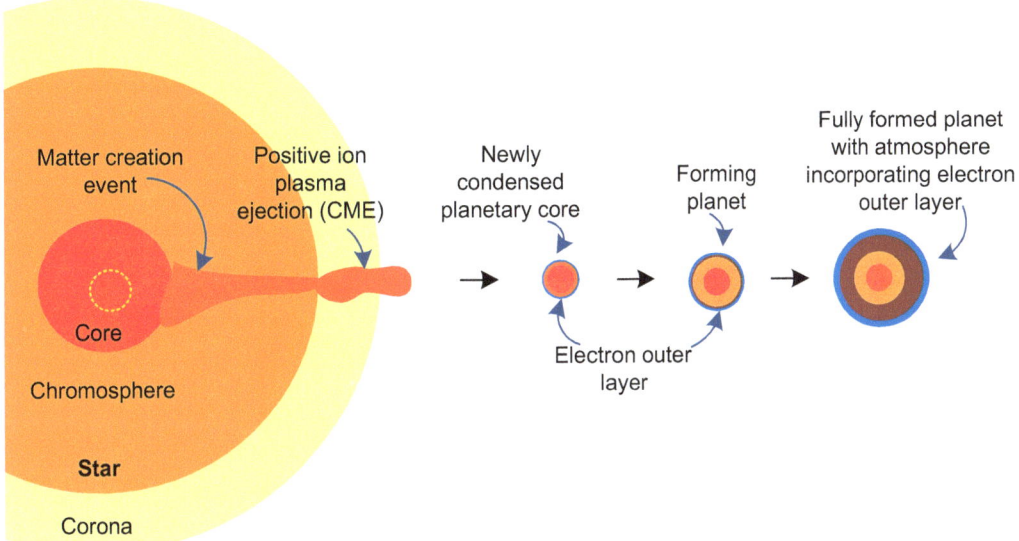

Figure 14.1. Planet X observation based planetary formation which agrees with Halton Arp's galaxy observations. A star ejects liquid plasma which condenses into a core which then ejects its own material so that the outer layers of the planet form [2].

The energy is released from within particles as free photons, they are then converted to matter, or particles and the excess goes back into the in-particle form of gravitational energy, which can also be described in terms of heat. Energy in this form can also flow from particle to particle through collisions. Particles with more of this energy in them will tend to vibrate and move faster which is equated to heat. The highest vibrating particles are the hottest and they will have the most gravitational energy in them. This is why magma is hot and hotter the deeper it is inside the earth. When particles collide, the heat from the most energetic particle flows to the one with least, so that they both end up with the same amount. I call this the principle of equal energy sharing.

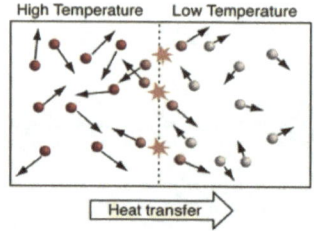

Figure 14.2. Heat is gravitational photon energy and manifests in the form of speed of a particle. It transfers to other particles through collisions, when matter is in the liquid and gaseous phases and through vibrational contact, i.e. conduction when matter is in the solid phase (outer solidified layers of a planet).

The energy released by the core of an established planet will go into the emission of infrared radiation by the surface of the planet. The energy release will reach a steady state, in which all the layers are kept at the same energy level or temperature (constant for each layer but differing across the layers with the inner layers being hotter), and a continuous and constant amount of energy is given off by the planet. An established star would also maintain its generation of energy at a steady pace in order to maintain its chromospheric layers, at a constant temperature, but it will also continuously give off matter as well as light. The matter given off is its solar wind and the light will be given off by its atmosphere which is kept in plasma arc mode, by the extremely high amount of gravitational photon energy in the electrons in its outer atmospheric layer.

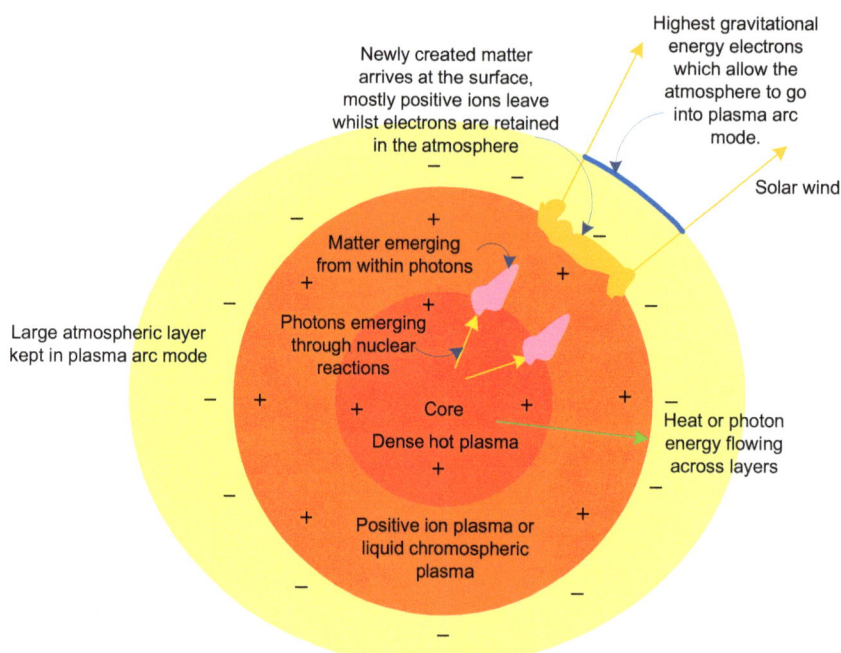

Figure 14.3. A star generates energy through nuclear reactions in the core such as converting protons to neutrons and radioactive decay which releases free photons. Free photons emerging into the high electric field split into particles which go into the production of star wind and renewal of electrons in

outer layer which are lost with the Solar Wind. These electrons allow the atmosphere to go into plasma arc mode and thus the emission of large amounts of light.

Energy in the core is generated through a more energetic means than through collisions and conduction as the high energy and density state of the core results in nuclear reactions. Particles release photons when neighboring particles have lower energy. But the release of photons sometimes lowers the gravitational energy of a proton or an electron to the point that they are no longer strongly enough repelled by each other through the charge separation part of the gravitational interaction, and are thus brought together by the electrostatic interaction, which turns them into a neutron. Some of the energy is converted to mass and some are given off as free photons so that the resulting particle has the same energy as other particles in its environment.

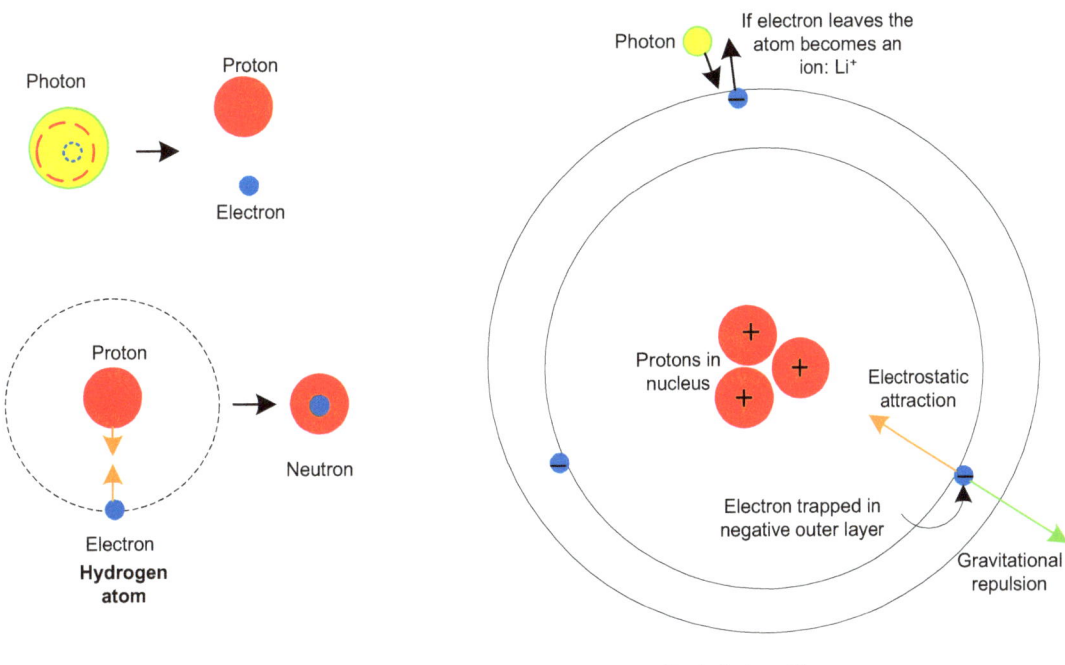

Figure 14.4. Photons split into their constituent particles. The gravitational interaction separates protons and electrons and attracts protons to protons so that neutral atoms can form. If the gravitational interaction, due to low photon energy onside the particles, is not strong enough, protons and electrons can combine into neutrons, in which case some of the gravitational energy gets converted to mass, and the rest excess energy is released as free photons. Electrons are trapped in an outer region of an atom, where the forces, exerted on the electron, due to the electrostatic and gravitational interactions, are in balance. We can see that electrons are energy within particles because when electrons absorb photons they are more strongly repelled by the nucleus so that they move to higher energy levels or leave the atom entirely.

In addition, some protons and neutrons may turn into neutrinos where now most of the mass is also turned into photon energy because neutrinos have either no mass or extremely low mass. This suggests that neutrinos are essentially a very low mass version of the neutron, which because of that extremely low mass do not interact through the strong force part of the gravitational interaction.

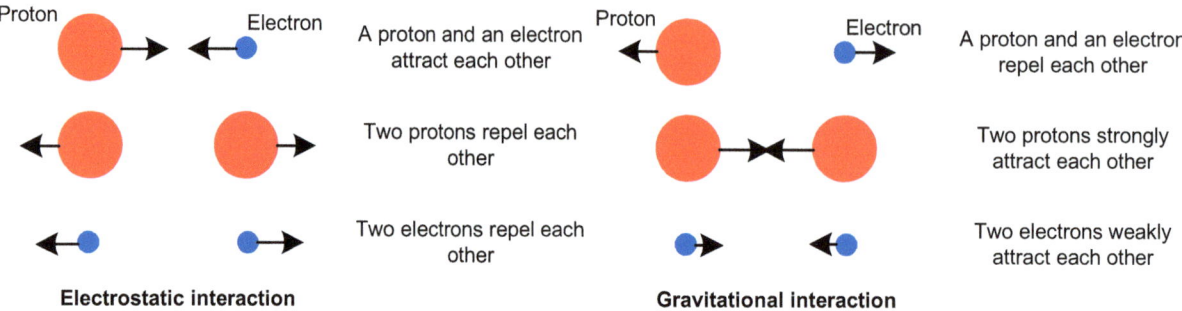

Figure 14.5. The electrostatic and the gravitational interactions between protons and electrons (see Book 3: Planet X Revealed Gravity and Light) [3].

A star's internal electric field drops, when protons become neutrons or neutrinos. Thus, over time, the core of a star will be converted into neutrons and its internal electrical field will decline, which mean that a star will end up as a neutron star. When a star becomes a neutron star, i.e. most of its protons have combined with electrons to form neutrons, it will no longer be able to create mass or emit light; it would have died. The dead star would also be very low in gravitational energy and thus have a very low ability to interact gravitationally and will, therefore, have a very weak gravitational field. Hence, a neutron star is a dead star, it will not give off the light, and it will only generate extremely weak gravitational and electric fields.

Planet X System Stellar Cores, however, are dead or energy depleted but are not neutron stars which agree with the idea that they died suddenly when energy suddenly departed from the core (see Article 513: Planet X planets are exploded planets) [4]; the loss in energy would cause the core and magma to become cold and solidify and also cause the core's gravitational influence over the outer layers to suddenly drop. These outer layers would be under severe pressure, due to being pulled inward, toward the core, so now, with the inward force gone, they would explode outwards and break into pieces, turning into a huge debris field of rocky pieces and dust. This would occur with both stars and planets.

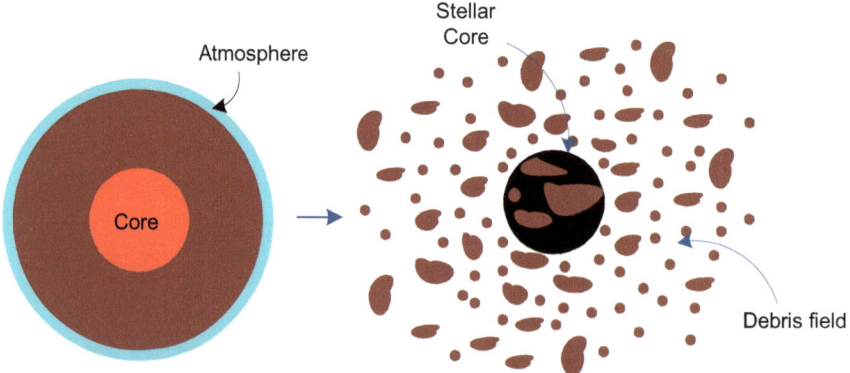

Figure 14.6. A living celestial object turned into a Planet X System Stellar Core when its gravitational photon energy suddenly was removed causing the liquid layers to solidify, all atmosphere to be lost and most of the gravitational field to be lost too so that outer layers explode outwards and break into pieces. The object will thus be made up of its original core and will be surrounded in a debris field.

The atmospheric matter and thus electrons in the atmosphere would not hold on to the debris pieces and would thus escape into space and be lost, leaving a positively charged core and debris field behind, although some pieces that were a part of the surface would remain neutral. The liquid water turned into tiny droplets, or clouds, and was attracted via the electrostatic interaction to positively charged pieces of debris and the core. In the core, the loss of gravitational photon energy would cause as many protons and electrons as possible to combine into neutrons or neutrinos, and many radioactive nuclei to quickly decay, which would release gravitational energy, which would then be shared through the core. This would be the core's last reserve of energy, which would allow it to retain a very weak gravitational influence and field. With protons combining with electrons to form neutrons or neutrinos, there would be a loss of positive electric charge and thus electric field but since a huge number of electrons were lost, because they were a part of the atmosphere, there would be plenty of protons left to continue to generate an electric field, which would thus remain much stronger than the gravitational field.

It is this electric field which allows a Planet X System Stellar Core to induce a matter creation event in the core of a living Solar System celestial object, which then results in a volcanic eruption or earthquake in the case of the earth or a CME in the case of the Sun.

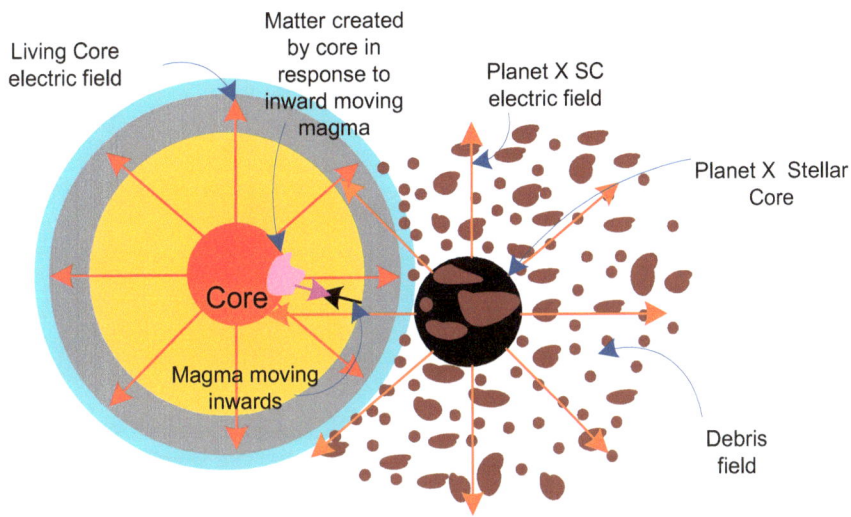

Figure 14.7. A living celestial object's positive field ends at the surface but a Planet X Stellar Core's field does not end, as the object has no negative outer layer, which allows it to enter the atmosphere of a living celestial object. Then, if the object is as large as the living celestial object's core or larger, as its electric field overlaps the living object's field, an instability occurs, which results in a matter creation event: the core releases a large amount of gravitational energy, which gets converted into matter.

The Planet X object's positive electric field causes positively charged magma in the earth's inner layers to either move inwards toward the core, in the opposite direction in which it normally moves, or it will slow down its normal motion, which causes energy in the core to build up, resulting in an instability, which then results in a very large release of gravitational photon energy as high energy free photons, which then split into particles as a result of the high electric field environment, this causes an explosion of newly created matter moving outwards toward the surface, in the direction of the Stellar Core.

In a newly formed planet, star or galaxy, what causes the matter creation is the fact that the core has too much energy for its size as it will have close to the same amount as the core of its parent but it will be much smaller, a large amount of gravitational energy for its size causes instability. In other words, there is an in the balance between the amount of energy being released by the outer layers and the energy inside the core so that matter creation is triggered until a balance is achieved. This causes the size and mass of the new celestial object to increase. Once it reaches the appropriate mass and size, for the size of its core, the energy generation will slow down to a maintenance level, where a steady state in energy flow from the inside is reached.

In conclusion, it has now become possible to understand the energy conversion mechanism leading to matter creation events, in operation inside the cores of planets and stars, in terms of the gravitational interaction theory, which I have developed and is based on Planet X observations. Matter creation occurs as a result of the normal evolution of a celestial object, be it a planet, a star or a galaxy, as well as in response to a Planet X System object, closely approaching a celestial object, when the Planet X System object is larger, or of a comparable size to the core of that celestial object. In order to support these energy conversions, the core of a living celestial object must be in a liquid plasma form, whilst Planet X System cores are dead and seem to be in a solid form.

References:

[1]   Albers, C. (2019). Article 541: Planet X cover-up and planetary formation: where do asteroids come from?
[2]   Albers, C. (2019). Article 532: Planet X or comet reenergizing process.
[3]   Albers, C. and C'one, S. (2018). Book 3: Planet X Revealed Gravity and Light.
[4]   Albers, C. (2019). Article 513: Planet X planets are exploded planets.

Chapter 15

558. Astronomical Quantum Mechanics: Gravitational potential and orbitals

In Article 555: Planet X and why Physics has failed? [1] I talked about how the Planet X observations have led to the understanding that all energy and matter come from light and how everything from atoms to stars, and galaxies, work according to the same patterns and laws. Thus, since stars operate like atoms, and may thus be called superatoms, and since atoms have energy levels, we would expect energy levels to also manifest at the astronomical level, i.e. we would expect stars and planets to also have energy levels. They should have energy levels on the inside so that energy decreases from the core to the top of the atmosphere, for all neutral atoms and ions making up each of the layers. Free electrons since they are negatively charged have increased energy the further you get from the core.

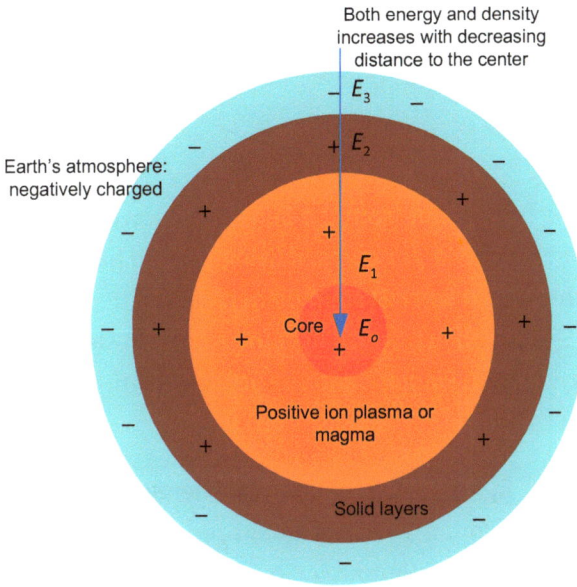

Figure 15.1. Gravitational energy per unit mass, or potential, increases with depth and is in the form of energy levels, matter in the core has the highest potential, matter in the molten rock region is in the next energy level, matter in the solid rock layer is lower in gravitational potential, and matter in the atmosphere is in the last and lowest energy level (see Article 514: Stellar Cores are gravitational poles or super proton) [2].

If a celestial object orbits another, it should orbit at a distance which is associated with its gravitational potential, thus celestial objects should have external energy levels as well as internal ones. Therefore planets orbiting the Sun should be in an orbit, which is associated with its gravitational energy. This is how it can be seen that Jupiter is an acquired object, i.e. a Planet X object re-energized by the Sun but still having a very large but yet dead core which therefore is very low in gravitational potential.

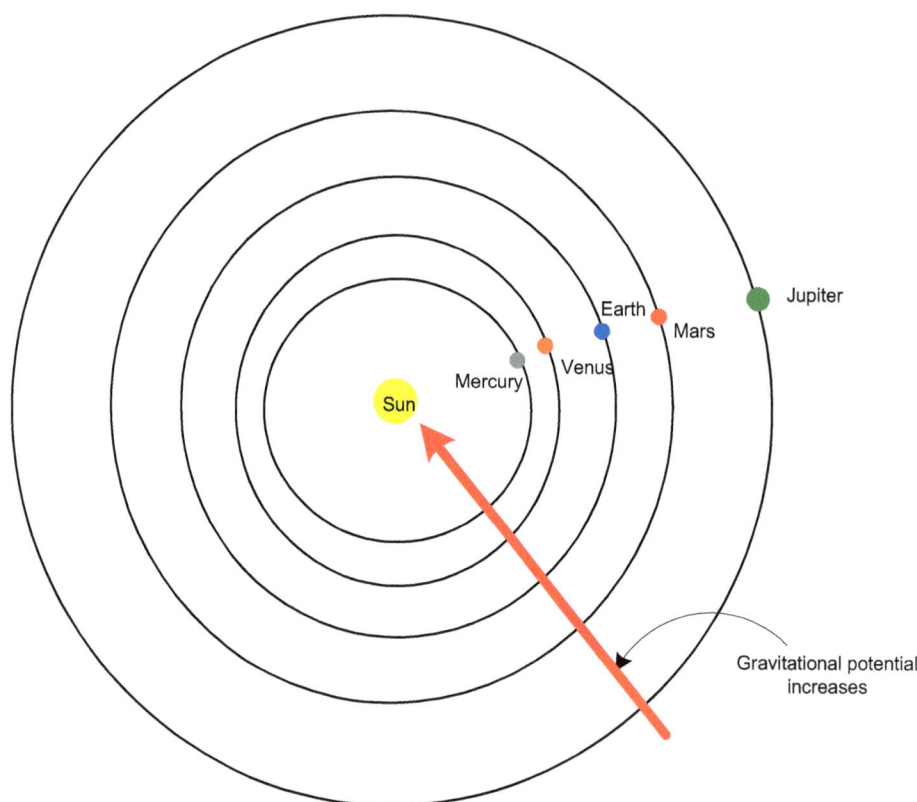

Figure 15.2. Planets orbiting further out from the Sun have a lower gravitational potential but larger objects should have larger cores and larger cores, if living, should have greater gravitational potential. This means that Jupiter, as the largest Solar System planet, cannot be a living planet; it must have a dead core and is thus a Planet X System object, which became a Solar System planet. The same applies to all the gas giant planets (see Article 523: Planet X and the Solar System: Jupiter and all gas giants are recent acquisitions) [3].

But, if all the planets in the Solar System orbit in different energy levels and we should be able to obtain an equation for gravitational potential, which is quantized. A quantized equation is one in which variables change in steps rather than continuously. So using known orbital parameters for Solar System objects, I am going to derive quantized equations, which describe the Sun's gravitational energy levels. I am going to use the relationship between the Sun's rotational period and the planets' orbital periods to come up with the required quantized equation and I would like to thank the subscriber who left a comment under one of my videos informing me of this relationship.

Now, the Sun's orbital period is not actually an easy thing to determine as it changes for different solar latitudes, it is about 24 days at the equator, but as fast as 36 days, at the poles. However, the movement of sunspots can be used to determine the Sun's average rotational period, which turns out to be between 27 and 28 days. I am going to use 28 days, as it is likely that the core moves a little faster than the outer layers of the sun, and because it divides exactly into Venus' and Earth's orbital periods.

Table 1. The orbital period for several planets in the Solar System is given in the 2nd column. The third column shows the result of dividing the orbital period of each planet, by the Sun's rotational period, T_{Sun} = 28 days. The 4th column shows the orbital quantum number for each planet, which is twice the number shown in the 3rd column. The last column shows the semi-major axis or average orbital distance from the Sun.

Planet	Orbital period: T_n (days)	Orbital quantum number: $n = T_n/T_{Sun}$	Energy level quantum number: $N = 2n$	Semi-major axis: a_n (au)
Mercury	88.0	3.14	6	0.387
Venus	224.7	8.02	16	0.72
Earth	365.2	13.0	26	1
Mars	687.0	24.5	49	1.52
Jupiter	4331	155	310	5.2
Pluto	90560	3234	6468	39.5

As can be seen from the table Earth's and Venus' orbital periods, when divided by the Sun's rotational period give almost exact whole numbers, but Mars' gives half a whole number, which suggests that the energy level quantum number is actually 2n. However, I will use n in the calculations because it simplifies them.

Mercury's period does not fit the scheme as well as Earth's and Venus', but Mercury seems to be a newly created planet and it is thus likely that it has not yet completely settled into a stable orbit, which may be why it has such a large eccentricity or elliptical orbit (see Article 524: Mercury: created by the Sun due to Planet X) [4]. Jupiter and all the gas giant planets, as new acquisitions are also not likely to have yet settled into their perfect stable orbits, so we can expect their orbital numbers to differ somewhat from whole numbers. Pluto, like Mercury, has a highly elliptical orbit and is therefore likely to have been recently created by the Sun, as well. So, according to the above scheme, Mercury is in the Sun's 6th energy level, Venus is in the 16th and Earth is in the 26th energy level. I am going to keep on using n, the orbital quantum number, though, which is 13 for the earth and 24.5 for Mars, to make the calculations simpler. So we have:

$$T_n = n T_{Sun} \tag{1}$$

And, Kepler's third law is:

$$T^2 = \frac{4\pi^2}{GM} a^3 \tag{2}$$

Substituting equation (1) into (2), we get:

$$(n T_{Sun})^2 = \frac{4\pi^2}{GM} a_n^3 \quad \Rightarrow \quad a_n^3 = \frac{GM}{4\pi^2} (n T_{Sun})^2 \tag{3}$$

Now, letting the gravitational potential, or gravitational potential energy per unit mass, associated with each energy level, be

$$V_n = \frac{GM}{a_n} \quad \Rightarrow \quad GM = V_n a_n \qquad (4)$$

Then, substituting equation (4) into equation (3), we get:

$$a_n^3 = \frac{GM}{4\pi^2}(nT_{Sun})^2 \quad \Rightarrow \quad a_n^3 = \frac{a_n V_n}{4\pi^2}(nT_{Sun})^2$$

$$V_n = \frac{4\pi^2}{n^2 T_{Sun}^2} a_n^2 \qquad n=1,2,3,.... \qquad (5)$$

Equation (5) is now a quantized equation for gravitational potential and thus gives the gravitational potential for each of the Sun's energy levels. The next step is to obtain a quantized equation for the orbital position. Using the known values for G (6.67 x 10⁻¹¹ Nm²/kg²) and the Sun's mass M (2.0 x 10³⁰ kg), we can find the first orbital radius, by letting n = 1 in equation (3):

$$a_1^3 = \frac{GM}{4\pi^2}(1T_{Sun})^2 \quad \Rightarrow \quad a_1 = 27\times 10^6 \text{ km} = 38.8 r_{Sun} = 0.18 \text{ au} \qquad (6)$$

where $r_{Sun} = 695\,508$ km. Thus, each orbital radii, orbital, can now be written in terms of the first orbital, or the orbital associated with the Sun's first energy level:

$$a_n^3 = n^2 a_1^3 \quad \Rightarrow \quad a_n = n^{2/3} a_1 \qquad n=1,2,3,.... \qquad (7)$$

To check that the formula is correct we will now try to obtain earth's orbital radius from it; earth's orbital quantum number is *n* = 13, so

$$a_{13} = 13^{2/3}(0.18 \text{ au}) = 1.0 \text{ au} \quad \text{or} \quad a_{13}^3 = 13^2(0.18 \text{ au})^3 = 1.0 \text{ au}^3$$

Now, substituting equation (7) into equation (5), we obtain an equation for the gravitational potential associated with each orbital quantum number or each of the Sun's energy levels:

$$V_n = \frac{4\pi^2}{n^2 T_{Sun}^2} n^{4/3} a_1^2 = \frac{4\pi^2}{n^{2/3} T_{sun}^2} a_1^2 \qquad n=1,2,3,.... \qquad (8)$$

Then, the Sun's first orbital would have the gravitational potential:

$$V_1 = \frac{4\pi^2}{1^{2/3}[28 \text{ d}(\frac{24 \text{ h}}{1 \text{ d}})(\frac{3600 \text{ s}}{1 \text{ h}})]^2}(27\times 10^6 \text{ km})^2 = 4917 \text{ km}^2/\text{s}^2 \qquad (9)$$

And all other gravitational potentials can now be written in terms of the gravitational potential, associated with the Sun's first orbital:

$$V_n = \frac{1}{n^{2/3}} V_1 \qquad n = 1, 2, 3, \ldots \qquad (10)$$

So that the Earth's gravitational potential is given by

$$V_{Earth} = V_{13} = \frac{1}{13^{2/3}} V_1 = 882 \text{ km}^2/\text{s}^2$$

Mercury's and Jupiter's gravitational potential would then be given by

$$V_{Mercury} = V_3 = \frac{1}{3^{2/3}} V_1 = \frac{1}{3^{2/3}} (4917 \text{ km}^2/\text{s}^2) = 2355 \text{ km}^2/\text{s}^2$$

$$V_{Jupiter} = V_{155} = \frac{1}{155^{2/3}} V_1 = 168 \text{ km}^2/\text{s}^2$$

Hence, the gravitational potential drops with increasing distance from the Sun as expected.

In conclusion, stars just like atoms have energy levels and thus using a relationship between the Sun's rotational period and the planets' orbital period, quantized equations for the gravitational potential and orbital radii of the planets have been obtained, as well as a scheme for determining the gravitational potential associated with each energy level.

References:

[1] Albers, C. (2019). Article 555: Planet X and why Physics has failed?
[2] Albers, C. (2018). Article 514: Stellar Cores are gravitational poles or super proton or white holes.
[3] Albers, C. (2018). Article 523: Planet X and the Solar System: Jupiter and all gas giants are recent acquisitions.
[4] Albers, C. (2018). Article 524: Mercury: created by the Sun due to Planet X.

Chapter 16

560. Planet X reveals that planets are smaller versions of their parent stars

In Article 555: Planet X and why Physics has failed? [1] I wrote about the fact that every effect has a cause and it is objects, Planet X Objects, in the Sun's environment that cause it to have CMEs and Solar Flares. These events are the result of matter creation events in the Sun's core. All celestial objects in the universe have cores, which create their own matter or outer layers, from the inside out, and also create new celestial objects, by ejecting core material (see Article 522: Stellar Cores are sources of matter or white holes) [2]. In Article 556: A mountain breaks apart in Siberia due to Planet X [3], I stated that the creating object has more energy than it can contain, which then causes it to create matter. This means that it has an outer layer of material that is several energy levels above that of its environment which then triggers matter creation.

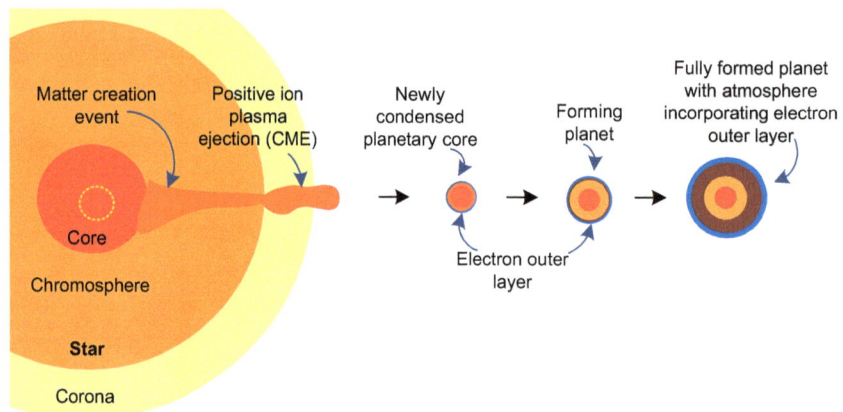

Figure 16.1. A star has a core ejection: it ejects part of its core inside the newly created plasma. The new core material once outside the body of the parent star ejects its own material so that it forms the outer layers of a new celestial object.

There seem to be two different types of matter creation, matter creation, which results in outer layers being formed and matter creation, which ejects matter to the outside of the creating object's body. This last type of matter creation ejects material further out and must be more energetic and thus contain higher energy material. The highest energy material available is its own core material but since it cannot create more material with the same energy as its core plasma, an ejection of the core material leaves a smaller core behind. The reason why it cannot create plasma with the same energy as the core's plasma is because some of the energy has to be converted into mass, thus lowering the energy which will be able to be inside the new particles. Newly created plasma is created from the release of photon energy in the core, which lowers its energy but no core plasma is lost. A core ejection allows a celestial object to create a smaller version of itself; so that planets are smaller versions of their parent stars. Only the highest energy CMEs will have core material in them, which can then lead to the formation of a new planet as seems to have occurred in the case of Mercury (see Article 524: Mercury: created by the Sun due to Planet X) [4]. Stars also continuously eject matter in the form of a Stellar Wind which gathers into nebular clouds centered around the star and forms the environment around the outside of the body of the star, in which any star core ejections will find themselves in. The material closer to the star will have more energy than the material further out.

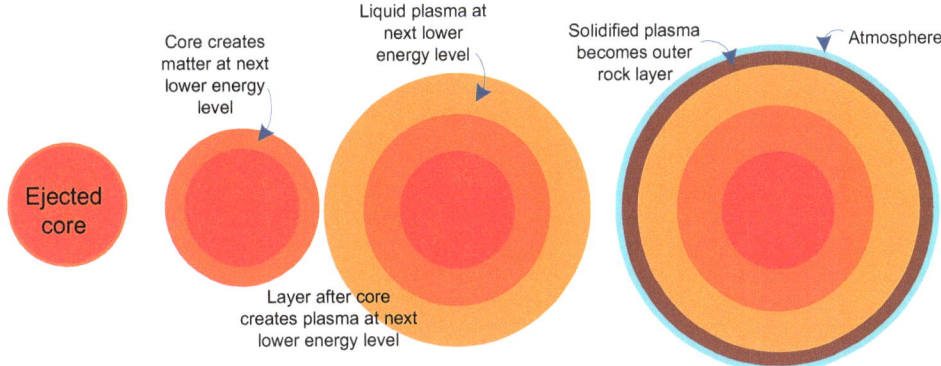

Figure 16.2. Ejected core material from a celestial object turns into a new celestial object: As the core is several energy levels above the environment, it ejects matter at the next lowest energy level, which in turn, ejects matter at the next lowest, until the outer layer is at a level just above the environment outside, which will be in the form of gases released from the liquid layers. Then the outer liquid plasma layer cools and solidifies, if the original core was small, i.e. it turns into a planet. Water vapor in the atmosphere condenses and turns into liquid water which collects on the rocky surface. But if the original core was large enough to create a high enough electric field in the outer layer, the outer gaseous layer will remain in the plasma state, or plasma arc mode and emit large amounts of light, in which case the outermost liquid layer will not solidify and the object will thus be a star.

In conclusion, Planet X Objects cause the Sun to have matter creation events which reveal that stars create planets which are smaller versions of themselves.

References:

[1]   Albers, C. (2019). Article 555: Planet X and why Physics has failed?
[2]   Albers, C. (2019). Article 522: Stellar Cores are sources of matter or white holes.
[3]   Albers, C. (2019). Article 556: A mountain breaks apart in Siberia due to Planet X.
[4]   Albers, C. (2018). Article 524: Mercury: created by the Sun due to Planet X.

Chapter 17

561. Planet X reveals how the universe began and how it is connected

In Article 560: Planet X reveals that planets are smaller versions of their parent stars [1], I detailed how celestial objects have two types of matter creation events, matter creation which increases the size of the object and matter creation, which ejects matter to the outside. The ejected material to the outside can have different levels of energy, with the highest level containing core material, which cannot be replenished, so, the core, which remains behind, will now be smaller in size. Galactic nuclei also have two types of ejections, ejections which create smaller versions of themselves, core ejections, which contain part of the galactic core, and turn into quasars, and lower energy ejections of newly created matter (no core material) which turn into globular clusters, break up into stars and thus populate the arms of a galaxy.

Figure 17.1. Left: Active Galactic Nuclei are intensely bright galactic centers, with extremely high electric fields. When the electric field and brightness reaches a critical level, it causes instability and the ejection of matter, which condenses into a quasar. In the course of time, the quasars start ejecting their own material, along with their major axis, which develops into arms and thus becomes galaxies themselves. The material ejected from the center of quasars spreads out in a spiral because of the rotational motion of the quasar. Thus galaxies give birth to galaxies, and matter is continuously being created in the universe. **Right:** A spiral galaxy, the spiral arms are due to the material being ejected along the galaxy's plane of rotation (see Article 126: White Holes instead of Black Holes at the Center of Galaxies) [2].

Thus, just as in stars, galaxies have matter creation events which result in the formation of another version of themselves (quasar) and matter creation events which goes into the increase of their size (globular clusters).

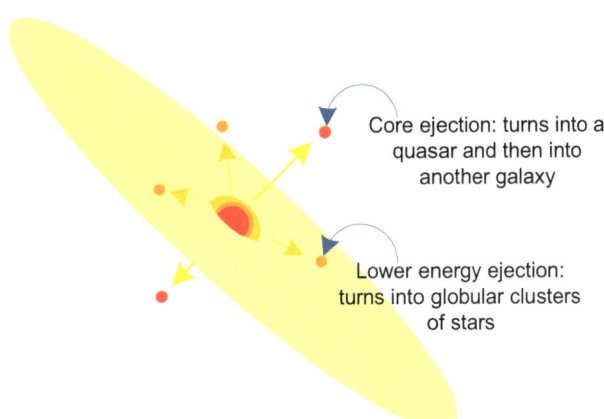

Figure 17.2. Galactic cores have two types of ejections: core ejections, which create smaller versions of the parent galactic core, i.e. lead to the formation of new galaxies and lower energy plasma ejections, which turn into globular clusters, or stars, which then populate the arms of the galaxy. The first type may contain core material surrounded by newly created plasma; the second type is made up only of newly created plasma.

The globular cluster material is lower in energy than the quasar material, so its core is lower in energy than a quasar's core, but matter creation and core ejection occurs in exactly the same way. The globular cluster core, now outside the body of its parent, encounters an outside environment at a much lower energy than it has, which triggers core ejections so that it splits into smaller pieces, and the smaller pieces split into even smaller pieces, until the cluster breaks up into individual stars. The continuously splitting cores also create outer layers and nebular clouds outside the body of each object, thus creating an environment that eventually their outer layers are in equilibrium with, thus stopping the core ejections and matter creation events. But as stars move away from each other, low energy regions appear which trigger core ejections. If a triggered core ejection is small, it turns into a planet.

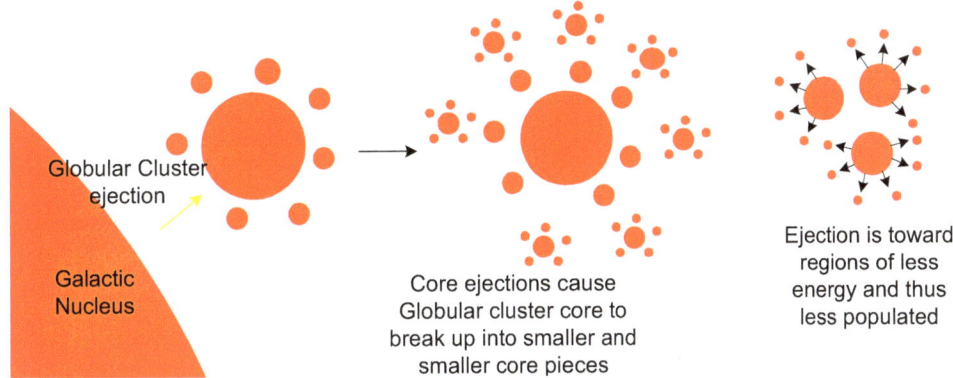

Figure 17.3. A Globular Cluster ejection turns into individual stars and spreads out from the central globular core, so that star density decreases with distance from the center.

Figure 17.4. NGC6388 Globular Cluster, in the Milky Way Galaxy: It is much denser at the center than at the edges indicating that it is the product of repeated core ejections spreading out from the central core. Ring patterns are apparent, where stars were ejected away from the same point, or parent core, at about the same time.

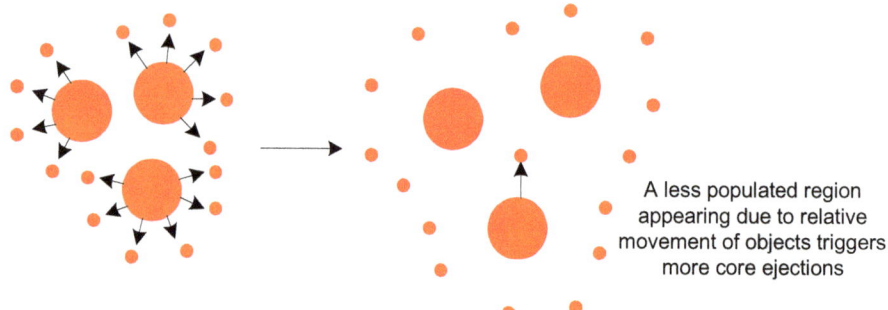

Figure 17.5. As breaking up cores, or forming stars, move away from each other, any region with a lesser population causes a drop in the energy of the local environment and triggers more ejections toward the less populated region. This is also what causes galaxies to have core or quasar ejections; as galaxies move away from each other, lower energy regions open up triggering new ejections, in the direction of the growing vacuum. Galaxies will tend to move away from the parent galaxy, and stars will tend to move away from the central globular cluster core, because ejections are always outwards, i.e. away from the parent.

Thus, all stars contain a piece of the parent galactic nucleus and all planets contain a piece of their parent stars so that all objects in a galaxy are connected to each other.

This mechanism also suggests how the universe began. It most likely began as a huge extremely high energy plasma core, which through core ejections broke into smaller and smaller pieces, eventually forming individual galaxies, and then galaxies did their own ejections which created more galaxies. As galaxies spread out, low population regions or energy vacuums appear, which trigger more core ejections from older galaxies, so that galaxies spread out and away from the original universal core.

In conclusion, matter creation and core ejections are occurring all over the universe. Celestial objects form themselves from the inside out, through matter creation, and create other celestial objects, which are smaller versions of themselves, through ejecting part of their cores. Planets are thus smaller versions of their parent star. This means that all planets are connected to their parent star and all-stars to their parent galactic nucleus. In addition, all galaxies are connected to their parent galaxies and all galaxies to possibly an original supergalactic core, so that the whole universe is connected.

References:

[1]   Albers, C. (2019). Article 560: Planet X reveals that planets are smaller versions of their parent stars.

[2]   Albers, C. (2018). Article 126: White Holes instead of Black Holes at the Center of Galaxies.

Chapter 18

563. Planet X creating sinkholes all over the world

Sinkholes have been appearing all over the world for many years now. Some are large and some are small but surprisingly a lot of them are circular which is very telling as to what is actually causing them. Figure 18.1 shows a sinkhole opening up in Brazil in 2014. Because of the houses falling into it and the angle from which the event is filmed, it is not clear whether the hole is circular or irregular in shape. However, too many sinkholes are perfectly round for it not to be a factor in us determining what is causing the sinkholes.

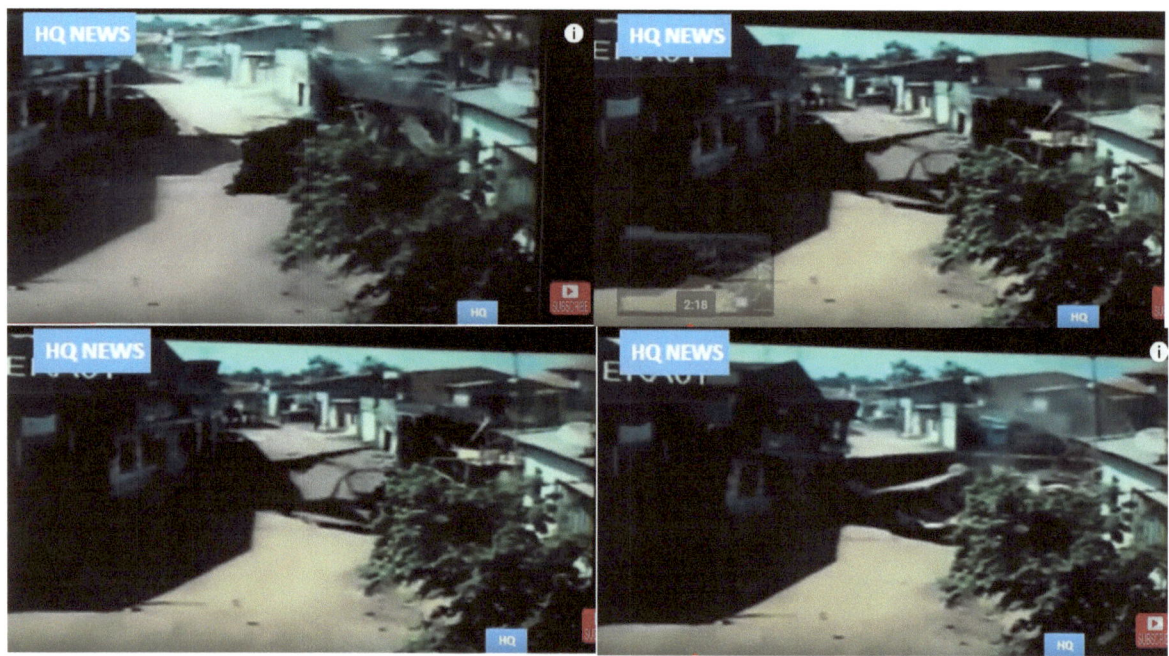

Figure 18.1. A large sinkhole opens up in Rio Brazil in 2014, destroying 3 houses.

Figure 18.2. This sinkhole which opened up overnight in the UK in 2014, is roughly circular on the outside but inside the walls of the hole seem to be even more circular than the opening at the surface suggesting the ground inside the hole was pushed down but the surface may have fallen in.

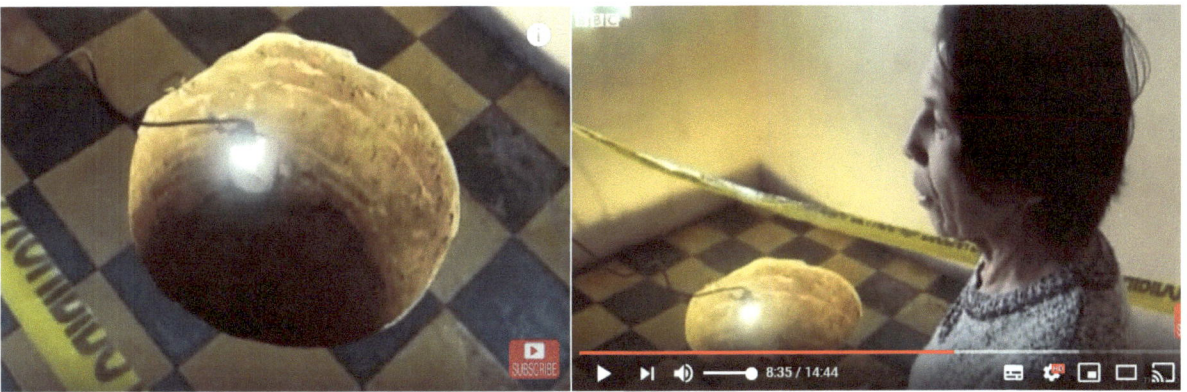

Figure 18.3. A small perfectly circular sinkhole opened up in this woman's bedroom, in Guatemala City, in July of 2011.

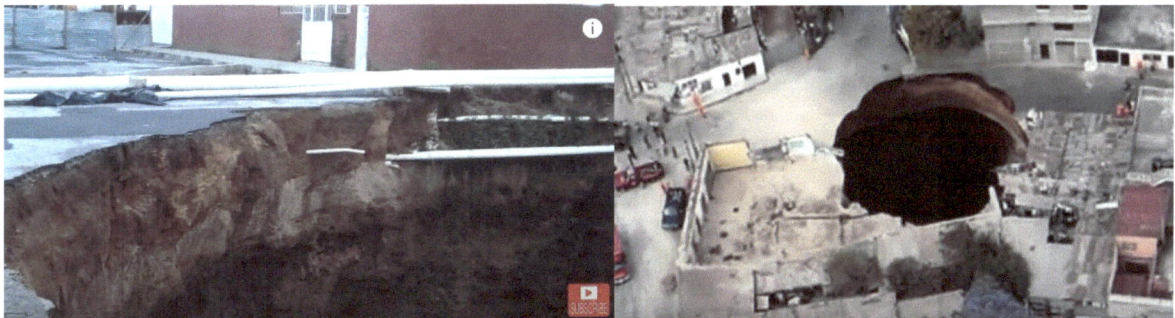

Figure 18.4. A large circular hole opened up in Guatemala in May of 2010.

If sinkholes were due to support underground collapsing through, for example, magma moving somewhere, leaving holes deep underground and thus the ground above losing support and collapsing, we would expect sinkholes to be irregular in shape, but they are clearly circular, instead. Circular sinkholes have to be caused by a force, which affects a circular region. Only a force exerted by a spherical object can affect a circular region. Thus, sinkholes are not due to something changing inside the earth, they are due to a force from above, a repelling force exerted by a spherical object above the ground.

Sinkholes form in many different sizes suggesting that either the objects exerting the forces come in different sizes. Sinkholes have different depths indicating that the repelling strength of the force varies; deeper holes are created by a stronger repelling force. The small but very deep hole in figure 3 shows that the sinkhole was created by a small object, which exerted a very strong repelling force for its size.

The fact that the repelling tidal force is exerted over such a small region suggests that it is tidal in nature, and thus exerted by objects very close to the surface of the earth. A tidal force is gravitational in nature, which shows that the gravitational force can cause repulsion as well as attraction. The only objects that can approach the earth close enough to give rise to such a sharp tidal force are, of course, Planet X Objects, or Planet X System Stellar Cores, which are the cores of dead stars, planets, and moons.

The fact that Planet X Objects can exert repelling gravitational forces provides an explanation for why ocean recession events occur.

Figure 18.5. Left: Ocean recedes leaving boats sitting on mud, in the harbor in Punta del Este, Uruguay, on August 11[th,] 2017. The ocean came back but this extreme low tide had never happened before. **Right**: An empty beach, due to the ocean receding, from the Brazilian coast, on August 12[th,] 2017, no large storms or hurricane could be blamed for the phenomenon, as there were no storms or hurricanes anywhere near this coastline. This too was unprecedented. Forces which affect the level of the ocean are tidal in nature and this shows that tidal forces can be repelling and thus cause low tides (see Article 227: Stellar Cores affecting the earth and possible connection to Volcanic Eruptions) [1].

In conclusion, Planet X Objects cause the formation of sinkholes when their tidal influence is repulsive. Planet X Objects' gravitational influence being repulsive also explains why ocean recession events occur.

Reference:

[1]   Albers, C. (2019). Article 227: Stellar Cores affecting the earth and possible connection to Volcanic Eruptions (in Book 6: Planet X Physicist Articles Part I).

Chapter 19

566. Planet X creating sinkholes and effects on the human body

In Article 563: Planet X creating sinkholes all over the world [1], I showed that sinkholes are created by Planet X objects when the gravitational interaction between the objects and the earth is repulsive. In Article 564: Sinkholes appear when gravity is repulsive [2] I showed that the gravitational interaction is repulsive when the objects are too low in gravitational energy, and thus sinkholes occur when the objects are new arrivals at earth. Hence, the appearance of sinkholes signals the arrival of new objects close to the earth.

Figure 19.1. Sinkhole: perfectly circular indicating that it was produced as a result of the gravitational interaction between the earth and a spherical object, Planet X Object or Planet X System Stellar Core, the dead core of a once living planet, above this point. Part of the surface can be seen lying in the sinkhole. The piece of paved road is in a large piece and thus appears to have broken and fallen in so it was not affected by the repulsive gravitational interaction like the ground below it.

The above image shows that the paved road was not affected in the same way as the ground below it. It shows that the acting force is not a pushing force from above; it is a force which acts on the individual particles of matter in the ground, not those in the road surface. It can also be understood in terms gravitational potential, the object's gravitational potential was too low for its altitude so the surface moved to accommodate it.

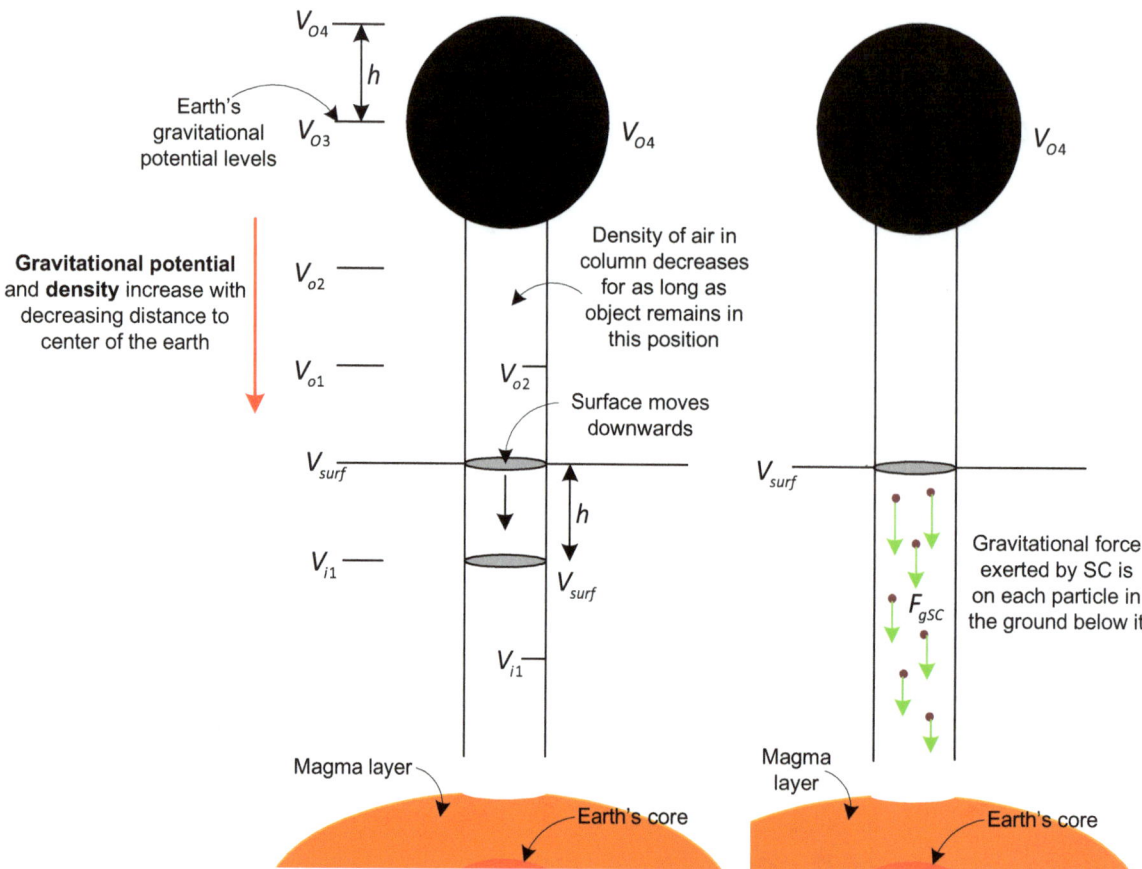

Figure 19.2. Left: Low energy Planet X Objects causes the earth's surface to move in order to accommodate its low gravitational potential which would result in the density of all matter at the same distance from the center of the earth, inside the column, to decrease. But the depth of the sinkhole is small with respect to the height of the atmosphere, so any difference in density and pressure will be extremely small. **Right:** The gravitational force acts on all particles making up the ground all the way to the magma layer. The force is strong enough to overcome the solid ground's resistance to breaking up. Since the whole column moves downward the force exerted by the Planet X Object extends to the magma layer. Thus, magma and the solid ground are matter at the right energy level and density for the sinkhole creating Planet X Objects to interact with. The road surface, atmosphere and people on the surface have too low an energy density to be affected by these objects. People and surface objects will simply fall into the sinkhole when it forms.

Thus, the sinkhole producing Planet X Objects do not interact with the atmosphere and cannot, therefore, create any real pressure changes, in the atmosphere. But, there are Planet X Objects which come into the earth's atmosphere and create severe low pressures, such as hurricane and tornado producing objects. This indicates that the objects are interacting with matter with a certain level of energy but since they are severely depleted in gravitational energy and potential, they can only be matching electrical energy as they retain a strong electric field and thus a strong electric potential. They would have lost some of their electric potential when they died but still retain a percentage, whilst they would have lost nearly all their gravitational potential. Thus, instead of gravitational tuning, we have electric tuning (see Article 576: Planet X larger than the Sun on a collision course with Earth: what happens?) [3]. Then, assuming all the objects retained the same percentage of their overall positive electric charge Q, their electric potential is given by

$$V_e = \frac{U_e}{q} = \frac{kQ}{r} = \frac{k\rho 4\pi r^3}{3r} \Rightarrow V_e \propto r^2$$

where r is the radius, q is a charge, ρ is charge density and k is the electrostatic constant, i.e. the electric potential is dependent on the radius squared. Thus, the larger the object, the larger its electric potential, and thus, the larger objects will interact with magma, deep inside the earth, whilst the small objects will interact with the atmosphere.

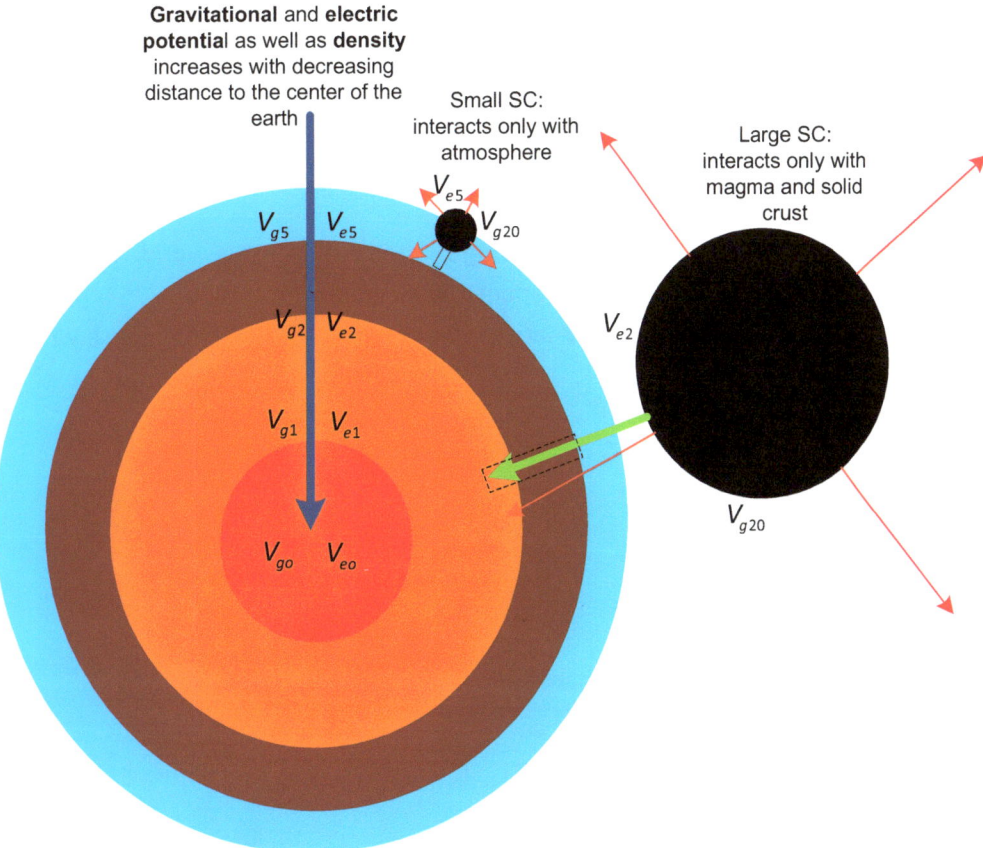

Figure 19.3. Planet X System Stellar Cores (SC) with a certain electric potential, interact with matter in the earth with the same electric potential. Large SCs interact with magma and the solid ground; the smaller SCs with the atmosphere. Medium sized SCs interact with liquid water. The objects only seem capable of absorbing energy from absorbing electrons through their cloud envelopes or by absorbing matter in the fluid phase, and thus from magma, liquid water or atmosphere. The objects are absorbing material with compatible levels of electric potential. They cannot be matching gravitational potential as they are severely depleted in gravitational energy, but their electric fields and electric potential although weaker than when they were living cores, remains strong.

People are therefore not likely to be affected by the larger Stellar Cores unless they induce matter creation events, which then produce earthquakes and volcanic eruptions. The smaller Stellar Cores which interact with the atmosphere and water, on the surface of the earth, are likely to affect the human body, however. As these objects move over a person, on the surface, air pressure drops, which may cause the ears to pop. The density of the blood, in a person's veins, will drop as well, which may cause the body to react; blood with less density will carry less oxygen and nutrients, which may result in a feeling of dizziness and weakness. The heart may pump harder in an attempt to compensate, which may lead to problems in people, with a weakened heart, and in people, with obstructions in their cardiovascular system.

In conclusion, Planet X System Stellar Cores of different sizes interact with matter of an electric potential, which is comparable to their own electric potential, so that the largest objects interact with only magma, inside the earth, and the smallest only with the atmosphere, whilst in between there will be those, which interact with liquid water on the surface of the earth. The smallest objects produce severe low pressures, which can have adverse effects on the human body but the larger ones have no effect on the human body because the human body has a much lower proton density than they do.

References:

[1]	Albers, C. (2019). Article 563: Planet X creating sinkholes all over the world.
[2]	Albers, C. (2018). Article 564: Sinkholes appear when gravity is repulsive.
[3]	Albers, C. (2018). Article 576: Planet X larger than the Sun on a collision course with Earth: what happens?

Chapter 20

569. Effect of Planet X Objects as large as the Sun on the Earth: energy levels

Planet X System Stellar Cores are the cores of dead stars, planets, and moons. They are dead because they are extremely depleted in gravitational photon energy, which is associated with the gravitational influence of a celestial object, i.e. its ability to generate a gravitational field, as well as its ability to create matter. All living cores from galactic nuclei cores to the cores of moons create their own outer layers of matter. Planet X cores are not able to. They do however retain some of their initial positive electric charge and thus some of their initial electric potentials. Since the dying process seems to have been sudden and would have had a similar effect on all the objects and since electric potential is proportional to the radius squared of the object, the larger objects would have a larger electric potential. The objects also interact only with matter of comparable electric potential to their own electric potential, so that the largest objects, of a size close to or larger than the Sun, find no material, even in the earth's core, with which they can interact, and will thus have no effect on planet earth. These larger objects that would once have been living stars, only find material that they can interact with on the Sun (see Article 566: Planet X creating sinkholes and effects on the human body) [1].

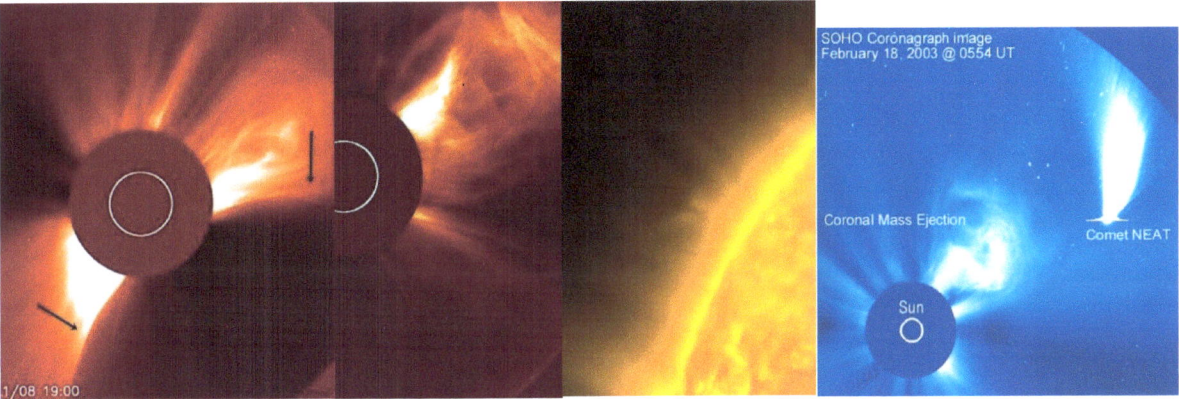

Figure 20.1. Planet X System Stellar Cores (SC) larger than the Sun, or of about the same size as the Sun, induce matter creation events in the Sun's core, which result in CMEs and Solar Flares. The smaller SC, in the yellow image, is only capable of interacting with the Sun's corona or perhaps the very tip of the chromosphere (Sun's liquid plasma layer). Objects from about 4 times larger than the earth seem to find compatible material to interact with, on the Sun. Smaller objects would have to go to the planets to find compatible materials they would be able to interact with. It is likely therefore that objects of about the same size as the earth, or perhaps up to twice the size of the earth, would be able to interact with earth's internal plasma, and induce earthquakes and volcanic eruptions, on earth (see Article 501: Planet X induced volcanic eruptions are like an Earth CME) [2].

This indicates that in a living celestial object, the electric and gravitational potentials have matching values and the core creates and interacts only with material which must be of the same energy level, which or one energy level below. It is also likely that an imbalance in the ratio, i.e. more gravitational potential than electric potential, triggers matter creation events. And a very large imbalance will likely trigger a core ejection, where a part of the core is ejected inside a layer of newly created plasma. This also suggests how Planet X Objects trigger matter creation events inside the cores of living objects such as the Sun or the Earth.

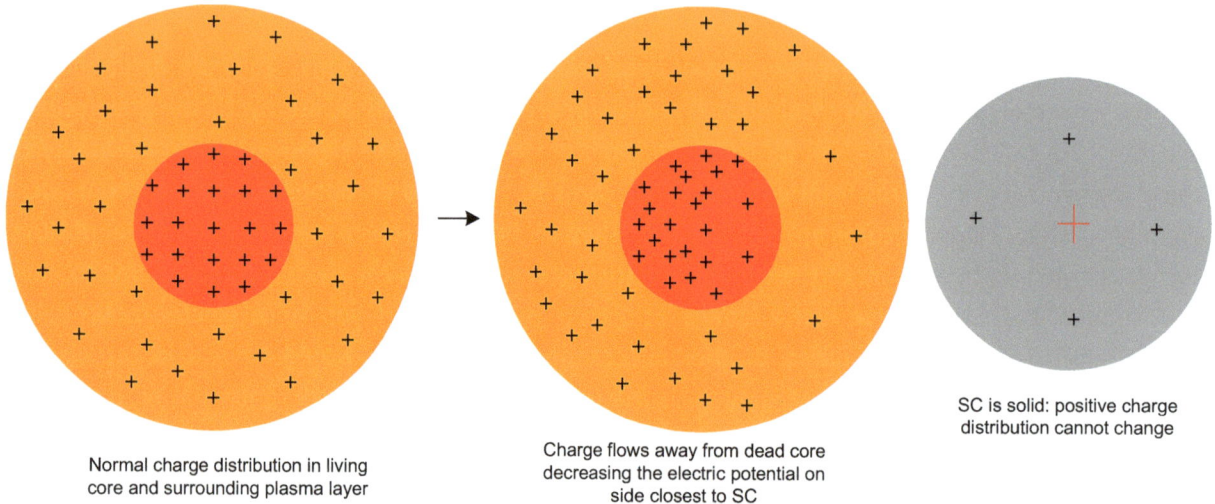

Normal charge distribution in living core and surrounding plasma layer

Charge flows away from dead core decreasing the electric potential on side closest to SC

SC is solid: positive charge distribution cannot change

Figure 20.2. Planet X Stellar Cores (SC) are severely depleted in electrons. When they lost their gravitational energy any remaining electrons combined with protons to form neutrons, the remaining protons remained protons because there were no electrons for them to combine with. It is these remaining protons that determine their electric potential and thus their energy level. But in the plasma of a living core, there are plenty of electrons to flow toward the region closest to the approaching SC which thus becomes more neutral, causing a drop in electric potential in that region. The electrons in the core are very low in gravitational energy, though (the higher energy electrons are repelled to the earth's outer negative layer) so the flow tends to decrease the electric potential but a greater than it is increased by the presence of the additional electrons (see Book 3: Planet X Revealed Gravity and Light) [3]. When the electric potential drops below the gravitational potential, a matter creation event is induced. If the SC is extremely large the drop in electric potential will be larger and a core ejection may be triggered.

The fact that SCs can only interact gravitationally and electrically with matter of about the same electric potential as they have suggests that both the electric and the gravitational interactions are energy level specific. It cannot just be the gravitational interaction that is energy level specific because then objects that cannot gravitationally interact with earth would still be able to trigger matter creation events. Thus, both interactions only occur between matter at the same energy level, as determined by its electric potential, and matter one energy level below. Therefore, the earth's core can attract its own core plasma and the plasma which its core creates, which we will name, inner mantle plasma. The inner mantle plasma can attract its own plasma and the plasma it creates, says, outer mantle plasma, but it cannot attract core plasma. This means that the highest potential plasma is always on the inside in a celestial object and there is no mixing of plasma at other energy levels. Since the electric potential is determined by the number of protons within matter, it is the number of protons per unit volume, which determines the energy level of matter, and what other matter it is able to interact with. Thus, the SCs as large as the Sun cannot interact with any matter found on earth because their proton density is much higher than any matter found on earth.

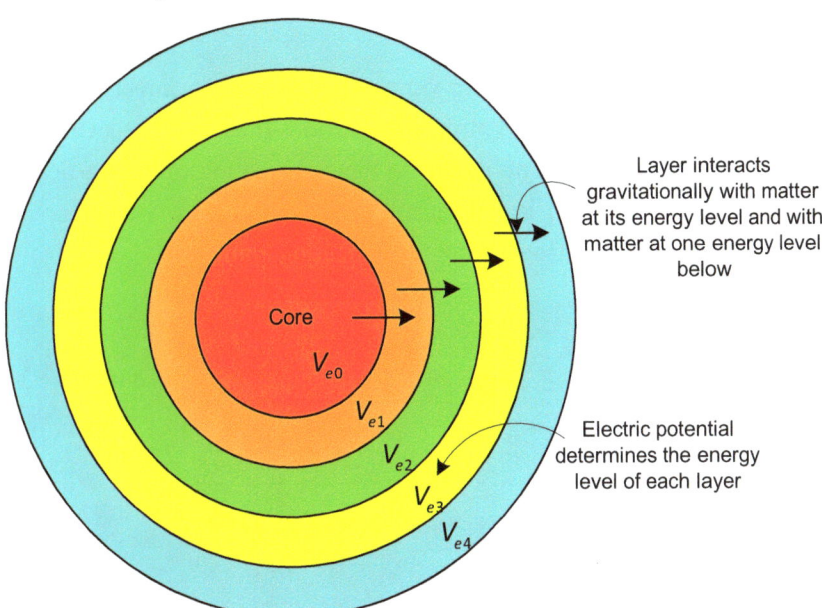

Figure 20.3. Matter within a layer has a characteristic potential which decreases outwards. Matter interacts gravitationally and electrically with other matter, at its own energy level, and with matter one energy level below. Because of the application of Newton's third law, we will have to add that matter of certain energy can interact with matter one energy level up. Since denser matter will have more positively charged particles per unit volume, the denser matter will have a higher electric potential and thus be at a higher energy level. This means that density will increase as we move toward the center or core of a celestial object.

In conclusion, the largest Planet X System Stellar Cores, those that are 4 times the size of the earth, or larger, can only interact gravitationally with the Sun, as their energy level, which is determined by their electric potential, which is in turn determined by their proton density, only allows them to interact with matter that is only found in the Sun. They, therefore, have no effect on the earth. Gravitational potential is determined by the amount of photon energy within the particles, making up each matter layer; both electric and gravitational potential increases toward the center of a planet. The gravitational potential usually matches (in living celestial objects) the electric potential. However, it can also be above or below. If the gravitational potential is above the electric potential, matter creation events, in which gravitational energy is released and transformed into matter (resulting in volcanic eruptions or earthquakes, on earth) are triggered.

References:

[1]   Albers, C. (2019). Article 566: Planet X creating sinkholes and effects on the human body.
[2]   Albers, C. (2019). Article 501: Planet X induced volcanic eruptions are like an Earth CME.
[2]   Albers, C. and C'one, S. (2018). Book 3: Planet X Revealed Gravity and Light.

Chapter 21

571. Planet X in our skies: water and vortices

Figure 1 below shows a web camera photograph showing a spherical object, in the sky, enveloped in cloud. The object is clearly inside the earth's atmosphere and is thus a Planet X Object, and from its size, which is small, it is most likely a small moon core. As I have shown in previous articles, Planet X Objects or Planet X System Stellar Cores (SCs) in the earth's atmosphere are always enveloped in a cloud envelope, made of tiny water droplets (see Article 570: Locating Planet X) [1].

Figure 21.1. Planet X Object in the Earth's atmosphere enveloped in its cloud envelope. The smallest of these objects seem to be about 250 feet in diameter (see Article 547: Planet X covered up by the moon hologram in the earth's atmosphere) [2].

The water making up a Planet X Object's cloud envelope is Planet X debris, as it used to be a part of the celestial objects, stars, planets, and moons, which died and turned into a dead core surrounded by a debris field; so the water, like all the rest of the material, including the core, are severely depleted in gravitational photon energy, or the energy, which is in the form of light, and that exists within particles. When these objects first enter the earth's atmosphere, the water in their cloud envelopes starts exchanging electrons, with the earth's atmosphere and thus starts to gain photon energy, which it passes on to the object. Since water will tend to equalize its gravitational energy to its environment, cloud at the bottom of the cloud envelope gains more energy than the top, which is why clouds are usually darker at the bottom. Eventually, enough of the cloud has gained enough energy to form a cloud vortex from the bottom of the object, which may eventually connect to the surface. The vortex once it reaches the ground level pulls material upwards along the rotating vortex. If the cloud vortex is over water it pulls water upwards toward the surface of the Stellar Core (SC).

Figure 21.2. Liquid water is transported upwards by this water spout. Water from the surface is denser and thus transparent whilst the Planet X water forming the cloud is not transparent, it looks opaque white, or some other color depending on whether it absorbs light or not. Some clouds emit light and are thus luminescent. This occurs when electrons settling in water molecule energy levels have too much energy and must release some in the form of visible light photons.

Now, as I detailed in Article 569: Effect of Planet X Objects as large as the Sun on the Earth: energy levels [3], matter only interacts with matter at a comparable energy level, which is determined by its electric potential and thus the number of protons per unit volume, in it, i.e. by its proton density. Larger objects have higher proton densities and are thus at higher energy levels. This results in smaller objects having comparable energy to matter in the earth's atmosphere and thus interact with the atmosphere, whereas larger objects are at higher energy levels and interact with plasma, inside the body of the earth. But since they all have water cloud envelopes, they are all able to interact with water, which indicates that water must be able to exist at vastly different densities and states. This is most likely why water has been found deep inside the earth. This water would be much hotter and denser than water on the surface and thus be in a different state, we may call it magma-water state. Water can also exist in the gaseous state in the earth's atmosphere but it will then be at low energy, the gravitational energy that was originally in the water having been used to break the bonds and thus cause a decrease in its electric potential and therefore its energy level.

Water in the cloud envelopes of Planet X Objects is in the liquid state but separated into tiny droplets, which means that its overall density is low, but the density in each droplet is high. This means that this water exists in two different states, at the correct overall low density, which is associated with its low initial gravitational energy, which is the same as the Stella Core's, but the droplets are at the correct high density, which is associated with the SC's electric potential or energy level. Because water adapts to its environment, inside the earth's atmosphere, it gains gravitational energy, causing its gravitational potential to increase and the water droplets to come closer to each other, i.e. the cloud's overall density increases, this tends to occur at the bottom of the cloud where the earth's gravitational potential is higher. Once the gravitational potential is higher than the electric potential, the electric potential jumps to the next electric potential or energy level, as it can only have discrete values, i.e. water has quantized electric potential (it varies in steps) but gravitational potential varies continuously, in water. When the water droplets increase their electric potential by one step, it is like a planet moving to a lower orbit, it moves at a higher speed, which is associated to that orbit. That starts the formation of a vortex at the bottom of the cloud.

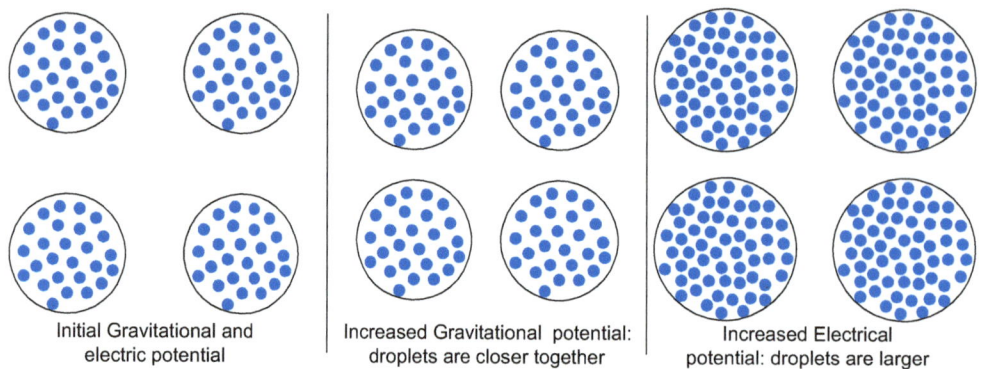

Figure 21.4. When water in SC cloud envelope gains gravitational energy the droplets get closer together and when the water jumps to a higher energy level, the droplets merge and larger droplets appear because the molecular bonds strengthen resulting in stronger surface tension.

Cloud vortices come in different sizes. The size of a vortex is associated with the size of the SC, the cloud is an envelope of. This is clear from the fact that size of water vortices, generated in a large amount of water, draining through a hole, is dependent on the size of the hole, and since the hole acts as a source of gravity, i.e. a massive object (see Article 380: Gravitational diffraction: gravity is a wave) [4].

Figure 21.5. Cloud vortices come in different sizes (radii or width) depending on the size of the Planet X Object generating them, the larger the object, the larger the cloud vortex. These clouds are usually darker at the bottom because water, at the bottom of the vortex, has higher gravitational energy and is, therefore, denser, i.e. the water droplets are closer together. The largest vortices seem to rotate slowly and the smallest much faster. **Right:** Surface water joins the cloud vortex and spirals into it, creating regions of lower than normal ocean level. The ocean water as it makes contact with vortex water changes its energy level to a higher level which causes the ocean level to drop.

We can also see that larger objects should give rise to larger vortices in terms of energy levels or orbits generated by an object of a certain size. Larger objects have larger electric potentials and will thus have energy levels or allowed orbits, which are further apart, with the first orbit having a much larger radius than the first orbit for a small object. This is why the stars like the Sun has much larger orbits than atoms, i.e. the Sun is much larger, so its energy levels are much further apart.

The fact that vortices with smaller radii rotate faster suggests that they are orbits, the water droplets are in orbit around a point at the center of the orbit. In the Solar System, the inner planets move much faster than the outer planets. For example, Mercury's orbital radius is 0.387 au and its orbital period is 88 days, so its orbital speed is

$$v_{Mer} = \frac{r_{Mer}}{T_{Mer}} = \frac{0.387 \text{ au}}{88 \text{ d}} = 0.0044 \text{ au/d},$$

whilst Earth has an orbital radius of 1 au and a period of 365 days, so its orbital speed is 0.0027 au/d, much less; thus, the larger the orbital radius the slower the speed. In other words, water droplets orbiting in a cloud vortex, are orbiting the central point in the vortex. The fact that they do not occupy the central region shows that this is not an allowed region, i.e. the energy level associated with an orbit, in that region, is not allowed, just like there are not allowed energy levels in atoms, where no electron can be found, and just like there will be certain orbital distances, in a planetary system, where no planets will be found (see Article 558: Astronomical Quantum Mechanics: Gravitational potential and orbitals) [5]. Thus, you would expect larger vortices, which are made of large orbits to rotate slower than smaller vortices, which are made of small orbits.

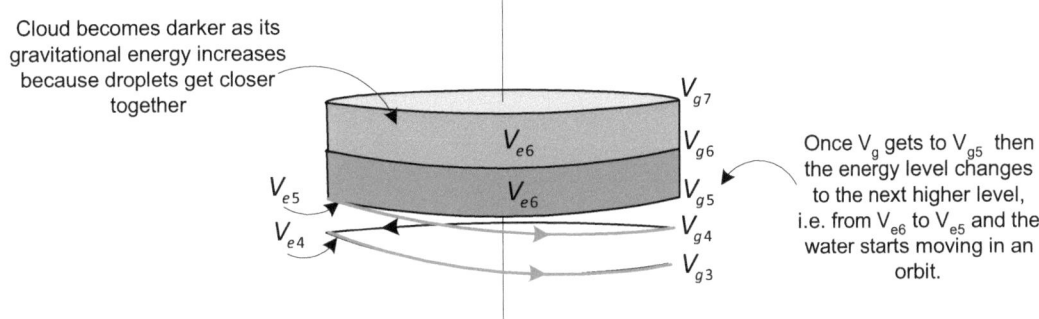

Figure 21.6. A cloud or water vortex: Gravitational and electric potential levels increase downwards: Gravitational energy of water changes continuously but the electric potential can only have discrete values. Once the gravitational potential is higher than the electric potential of the SC and thus higher than all the water in the cloud, then the water at the higher gravitational potential changes to the next higher energy level, which causes it to go into orbit about a central point at the level it has reached. The central point is in line with the center of the SC and the center of the earth. The size of the orbit depends on the size and electric potential of the SC. The orbital direction is determined by the local magnetic field, as the circular motion creates a magnetic field, which will then anti-align with a local magnetic field. If the local magnetic field is oriented downwards (Northern Hemisphere), the rotation will be as shown as it produces an upward oriented magnetic field (anti-clockwise).

In conclusion, Planet X System Stellar Cores, surrounded in their cloud envelopes, produce cloud vortices, in the earth's atmosphere, which provide a bridge between the Stellar Cores and the earth and allows them to absorb energy and matter from the earth. The formation and shape of the vortices can be understood in terms of energy levels, and associated orbits, as well as from the way that water as a substance is able to do what other substances cannot: it can continuously change its gravitational potential by varying the distance between droplets of water, whilst changing its electric potential in a step by step, or quantized, way.

References:

[1]   Albers, C. (2019). Article 570: Locating Planet X.
[2]   Albers, C. (2019). Article 547: Planet X covered up by the moon hologram in the earth's atmosphere.
[3]   Albers, C. (2019). Article 569: Effect of Planet X Objects as large as the Sun on the Earth: energy levels.
[4]   Albers, C. (2019). Article 380: Gravitational diffraction: gravity is a wave.
[5]   Albers, C. (2019). Article 558: Astronomical Quantum Mechanics: Gravitational potential and orbitals.

Chapter 22

573. Planet X reveals why everything rotates and what particles look like

Planet X System Stellar Cores come in the Solar System as comets and comets follow curved paths. In addition, newly created planets start out having rotational motion because their parent star is already rotating. But why does the parent rotate? This has to be due to particles themselves rotating.

Figure 22.1. The Andromeda Galaxy: The galactic core rotates, star clusters orbit the galactic core and large stars in star cluster will have smaller stars and planets orbiting them. Also, each star and planet will also rotate. Thus, all objects in a galaxy have both orbital and rotational motion.

Now, matter and thus particles appear when the core of a celestial object goes through a matter creation event. This occurs when the gravitational energy, in some region of the core, surpasses the electric potential, which results in the release of some gravitational energy (see Article 569: Effect of Planet X Objects as large as the Sun on the Earth: energy levels) [1]. Gravitational energy is photon energy, i.e. photons, which exist within particles, and thus the release of gravitational energy is the same as the release of photons from within particles, which thus emerge as free photons and immediately are converted to matter, or particles. The process is the same as what is described in accepted physics theory, as particle anti-particle creation, except that there is no such thing as an anti-particle, all we have is two particles of opposite charge, emerging from within a photon. The mass of the two particles depends on the energy of the photon, the particle pair only has the same mass when the photon energy is low, in which case, one of the particles is an electron and the other is a particle with the same mass but positively charged, called a positron. A more energetic photon (with the energy of about 1000 MeV) causes the particle pair to be an electron and a proton (see Book 3: Planet X reveals Gravity and Light and Article 196: Stellar Cores, particle mass, and replicator technology) [2, 3].

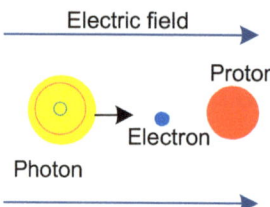

Figure 22.2. When a photon, released by the plasma core, inside a celestial object, and goes through a region of the high electric field, the two oppositely charged particles, inside it, are pulled apart causing the photon to split and the particles to move to the outside of the photon. The photon transforms into energy in the form of mass, for the two particles and what is left becomes gravitational energy shared equally by the two particles [2].

The particles must somehow start rotating, as well. Now, we know that all charged particles have magnetic moments, i.e. an intrinsic magnetic field, and since magnetic fields arise from charged particles moving along circular orbits, there must be a charge inside each particle moving in a circular orbit. This suggests that the charged particles, which emerge from inside the photon, start orbiting the photon, which is thus like the central star for the particle. I will call the massless particles: Negaton and positon.

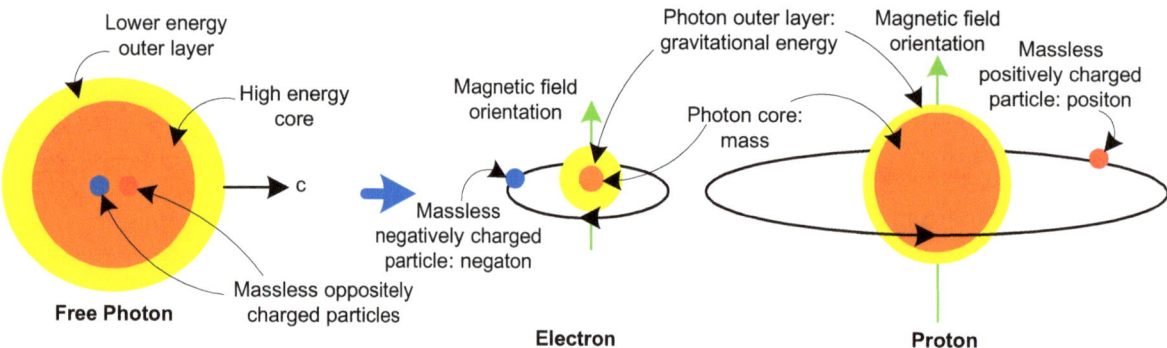

Figure 22.3. Particle structure: massive particles are like a simple star system. The charged particles (negaton and positon) start out inside a free photon moving at the speed of light and remain massless even outside the photon. The photon has a core and an outer layer. When the photon moves through a region of the high electric field, the charged particles, inside the photon, separate, causing the photon to split into two parts, and slowing its speed dramatically. The negaton and positon then orbit the two resulting photons, which thus become like central stars to them. Mass comes from the size of the photon core, at the center of the orbit. The outer photon layer is the particles' gravitational energy, which allows the gravitational interaction to take place. The negaton and positon remaining massless makes it possible for the two particles, resulting from a photon splitting, to have different masses depending on the energy or size of the original photon core, i.e. we could get an electron and a positron, an electron and a proton, or a proton and an antiproton, or any other hadron pair of opposite charge. The orbital motion of the charged particles gives rise to the electron's and proton's magnetic field.

Since the negaton and positon orbit the photon as if the photon is a central star, we would expect the photon to also rotate. Hence, the straight-line light speed of the photon and of the two massless charges is transformed into the orbital speed of the two charged particles, and rotational speed for the photon, at the center of each particle. Particles with less gravitational energy will tend to orbit those with more gravitational energy, thus resulting in celestial objects having higher gravitational energy centers and rotating about an axis, through their centers. Magnetic fields of particles will tend to also align thus building up to a large magnetic field for the whole celestial object.

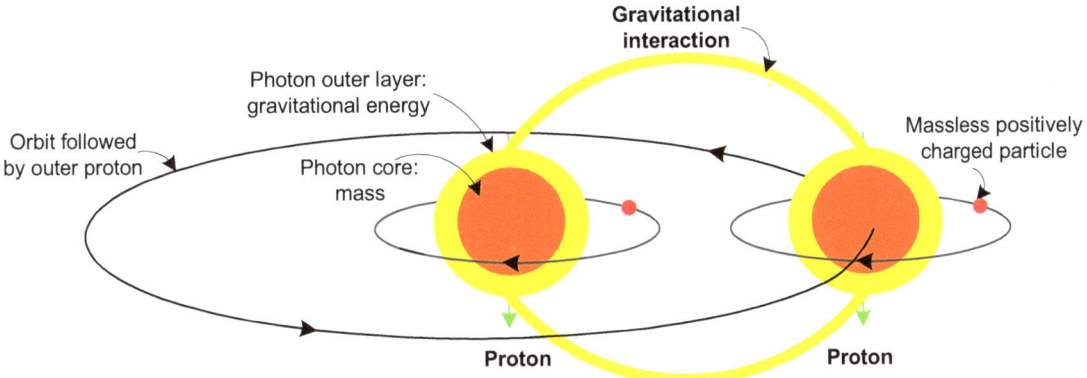

Figure 22.4. Two protons attract each other through the gravitational interaction: The photon at the center of each proton rotates and the protons orbit each other. The positive charges (positions) will repel each other through the electrostatic interaction stopping protons from ever colliding.

The photon core is associated with the mass of the particle (proton), which never changes and the outer layer is associated with the gravitational energy of the particle, which can change. The electric potential is given by the energy level or orbit of the charged particle. If the gravitational energy or potential is too high for the electric potential, i.e. the charged particle's orbit is too close to the photon star, it results in the release of gravitational energy, or of some of the photon outer layer.

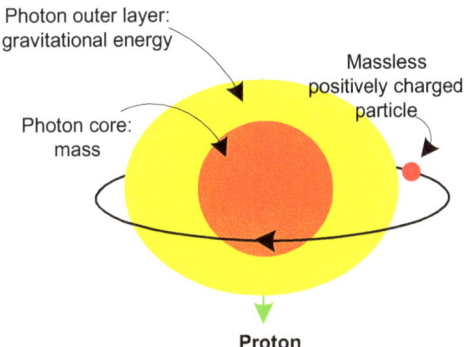

Figure 22.5. A particle's electric potential is set by how close the charged particle orbits the photon center, if the gravitational energy is too high, i.e. too close to the orbit; some of the gravitational energy is released. Released gravitational energy emerges from inside the particle as a free photon. Because photons are quantized, all energy releases, from particles, are also quantized, i.e. they are released in packets called quanta.

In conclusion, understanding, of how matter interacts through the gravitational and electrostatic interactions, arising out of Planet X observations, has now led to the understanding of how particles are structured, where photons are like a central star and a massless charged particle orbits it.

References:

[1] Albers, C. (2019). Article 569: Effect of Planet X Objects as large as the Sun on the Earth: energy levels.
[2] Albers, C. and C'one, S. (2018). Book 3: Planet X reveals Gravity and Light.
[3] Albers, C. (2018). Article 196: Stellar Cores, particle mass, and replicator technology.

Chapter 23

576. Planet X larger than the Sun on a collision course with Earth: what happens?

Planet X System Stellar Cores have been observed close to the Sun on many occasions. These objects come into the Solar System as comets (see Article 367: Planet X coming in as comets and affecting the Earth) [1] and must, therefore, enter the Solar System along extremely eccentric orbits, which must have taken them past the earth's orbit. Since comets have been observed for thousands of years, these objects must have been coming past earth's orbit for thousands of years, so it is important to consider what happens when one heads towards the earth.

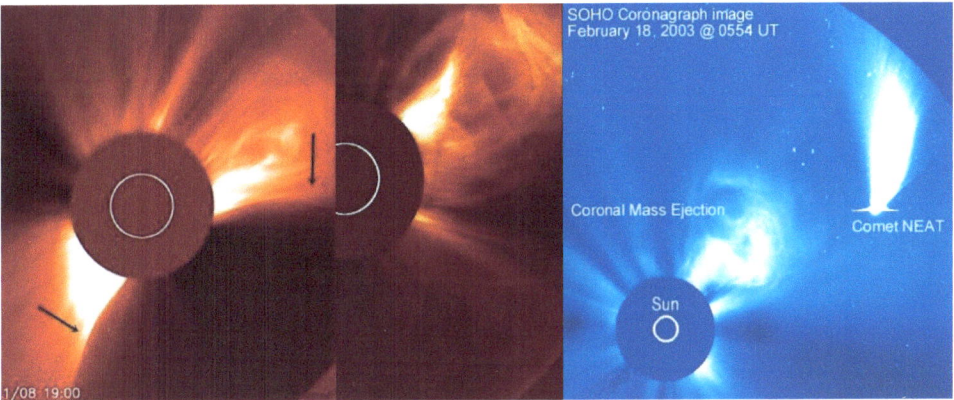

Figure 23.1. Planet X System Stellar Cores larger than, or about the same size as, the Sun, in close proximity to it; i.e. in the Sun's inner or outer corona. These objects induce matter creation events in the Sun's core, which result in CMEs (coronal mass ejections) (see Article 542: Planet X observation based energy conversion in the cores of planets and stars) [2]. The objects approach the Sun as comets.

Planet X System Stellar Cores approach the Sun along different orbits and only the largest objects have a large enough electric potential to get to the Sun. This is because the Sun has energy levels and an object at a certain energy level tries to get closer to the sun than the distance which is associated with that energy level, it is repelled. The greater the size of an object the greater will be its energy level associated with the object's electrical potential or proton density. So only the largest objects have a high enough energy level to reach the Sun (see Article 566: Planet X creating sinkholes and effects on the human body) [3].

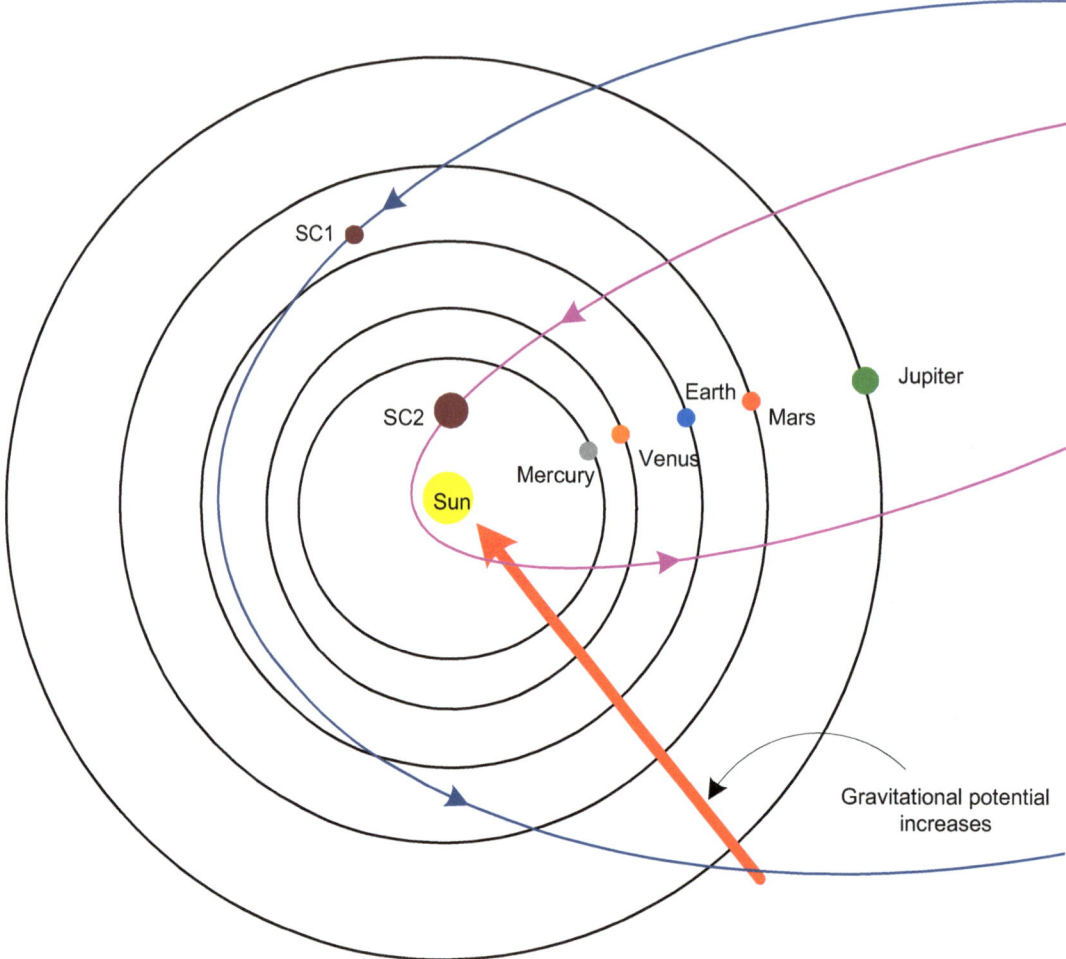

Figure 23.2. A Stellar Core (SC), with an average electric potential (energy level), which is just a little higher than earth's, will be able to approach the Sun to a distance, which is just inside the earth's orbit, and a SC with an electric potential close to the Sun's will be able to reach perihelion much closer to the Sun. The larger the object the higher its energy level, which shows that only the largest objects reach the Sun; in other words, Sun diving comets have to be very large Stellar Cores.

If a SC's perihelion position is inside the Sun's atmosphere, it will connect to the Sun, as if it is a part of it, for a period of time; it may also remain stationary, with respect to the Sun's surface, until a drop in electrons in the Sun's atmosphere results in it being ejected and going back to a long elliptical orbit (see Article 373: Planet X System orbits) [4].

In addition, matter at a certain proton density seems to only interact with matter of comparable proton density, so Stellar Cores of different sizes interact with matter at different depths, within the earth, which is why only the largest Stellar Cores (SCs) give rise to sinkholes (see Article 566: Planet X creating sinkholes and effects on the human body) [3].

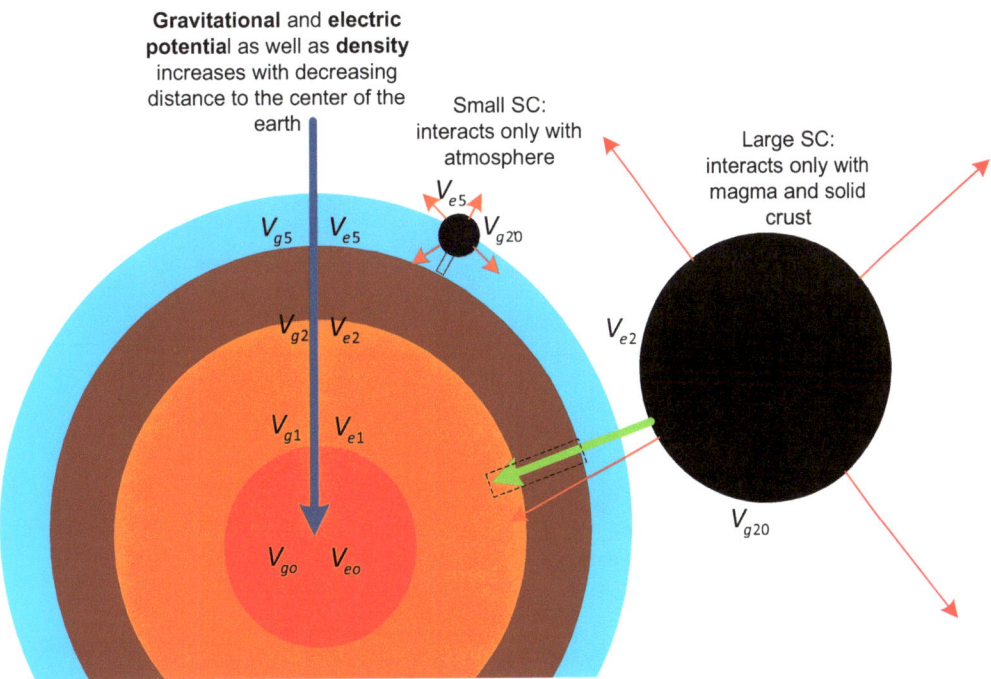

Figure 23.3. Celestial objects have internal energy levels which can be described in terms of their gravitational and electric potential: Stellar Cores (SCs) interact with matter, of the same proton density, it has, i.e. they interact with the matter layer at the same energy level as its own matter layers. So, if the SC has layers at different energy levels, matter at each energy level will interact with a layer, on the earth, at the same energy level.

Now, what will happen when one of these very large Stellar Cores is on a collision course with earth? These objects have at a much higher energy level than the earth, but yet it is impossible for them to pass right through the earth without affecting it, as both are solid objects. Since celestial objects tend to repel each other just like protons, in the nucleus, repel each other, it is to be expected that the two objects will repel each other before a collision occurs, which indicates that even though the earth is at a much lower average energy level, than the very large Stellar Core there must be some matter within it, which is at the same electric potential as some of the matter in the Stellar Core. This means that there has to be matter, within the earth, that is at the same electrical potential and thus proton density (energy level), as the matter in the Stellar Core, and since the same argument would arise with any object, no matter how large, on a collision course with earth, this suggests that within the core, of all celestial objects, remains a layer deep inside, which is at the highest energy level possible, in the universe. In other words, each core retains a piece of the original supercore that all celestial objects, in the universe, came from (see Article 561: Planet X reveals how the universe began and how it is connected) [5]. This central maximum energy level matter, at the center of a core, will be smaller, in smaller objects, but there is still some material at the original electric potential and thus at the original proton density, of the original supercore.

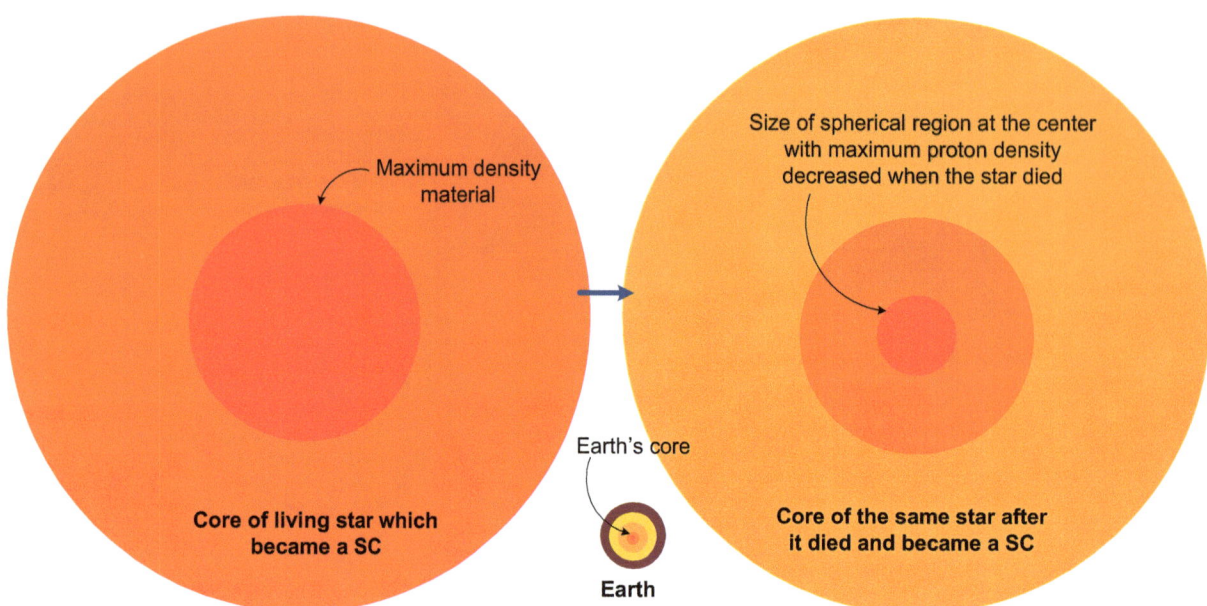

Figure 23.4. The core of a large star would have a large central region at a maximum proton density, represented by the red color, and another layer around it at a lower proton density. The core of the same star had a much smaller central region at the maximum density, after it died because electrons would combine with protons to form neutrons, when the star's gravitational energy was removed, i.e. when it died. The earth also has inner layers at the same maximum density and will thus be able to interact with the star. However, the star does not have any layers of the same density as the earth's outer layers and will thus not be able to interact with them. Small SCs, however, which seem to only interact with earth's outer layers may have become so depleted that they have lost all or almost all the inner layer at the maximum density and thus have no effect on the earth's inner layers and cannot, therefore, induce matter creation events or attract magma once it exits through a volcanic eruption.

This also makes sense in terms of creation of celestial objects, such as planets, which form inside stars and then are ejected. In order for the star and the new planet to remain connected and thus interacting with each other, the new planetary core has to retain a piece of the parent star, at the same electric potential or proton density, as the parent's own core. The new planetary core will thus use its energy to create new matter, but it will always retain a piece, inside it, that is just as dense and as energetic as its parent star's center.

Thus, the earth will interact with all size Stellar Cores, although the very large objects will not be able to get much in terms of energy or material, from such a tiny planet as earth, and will thus tend to go to the Sun, for energy and electrons. Since there is matter, within the earth, that will react to the presence of such an object, if any pass close to the earth, matter creation events will be induced deep within the earth, which will result in volcanic eruptions or earthquakes. The earth will also most likely be tossed around like a ping pong ball, as the object passes close to it. However, the earth will return to its orbit after the object passes, since the earth's orbit, as well as any other celestial object's orbit, is determined by its average energy level.

Since the Sun is no longer shining due to its current state of electron depletion it is also likely that more and more, as well as larger and larger SCs, will find their way to the earth with cataclysmic consequences for earth (see Article 535: The Sun is no longer shining: what has happened to it?) [6]

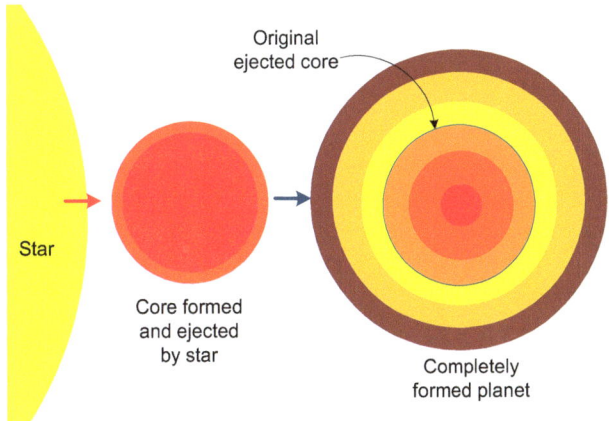

Figure 23.5. A core is created by a star and ejected. The core once outside the parent turns into a complete planet by creating outer layers of material at lower proton densities (see Article 560: Planet X reveals that planets are smaller versions of their parent stars) [7]. In the process, the star uses some of its gravitational energy, which causes its proton density to decrease, as when gravitational energy decreases, protons, and electrons combine to form neutrons, which causes a drop in the proton density. Thus, the size of the highest density regions in the core decreases. The average proton density, and thus, the electron potential, of the object, drop, as its average gravitational energy also drops, as the core creates its outer layers.

In conclusion, all living celestial object cores have at the center matter which is as dense and energetic as the plasma at the center of the original supercore in the universe from which all celestial objects come from. This is the only way to avoid objects colliding with each other, a very large Stellar Core on a collision course with earth will be repelled by the earth, and the earth will be repelled by it so that no collision is possible. However, the object will only be able to interact with the earth's core and have no effect on the earth's top layers. The approach of such large SCs to earth is likely to lead to cataclysmic volcanic eruptions and earthquakes.

References:

[1]   Albers, C. (2018). Article 367: Planet X coming in as comets and affecting the Earth.
[2]   Albers, C. (2019). Article 542: Planet X observation based energy conversion in the cores of planets and stars.
[3]   Albers, C. (2019). Article 566: Planet X creating sinkholes and effects on the human body.
[4]   Albers, C. (2018). Article 373: Planet X System orbits.
[5]   Albers, C. (2019). Article 561: Planet X reveals how the universe began and how it is connected.

[6]   Albers, C. (2019). Article 535: The Sun is no longer shining: what has happened to it?
[7]   Albers, C. (2019). Article 560: Planet X reveals that planets are smaller versions of their parent stars.

Chapter 24

577. Planet X created the asteroid belt and rocks in the sky

Planet X System Stellar Cores have been coming into the Solar System for thousands of years and come in surrounded by a debris field, which is made up of rocky pieces and water, which is in the form of clouds (see Article 575: Planet X causes atmospheric electron depletion, which is life threatening) [1].

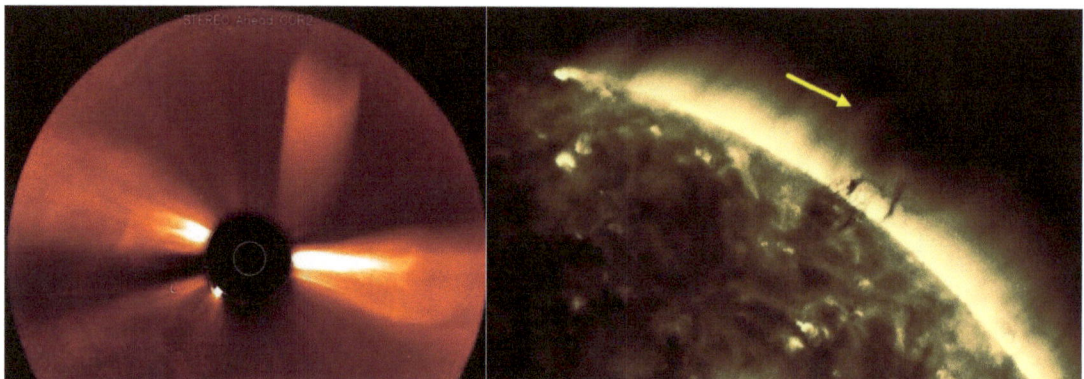

Figure 24.1. Planet X System Stellar Cores in the Solar System: The one on the left is moving close to the Sun and is leaving a trail behind it. The one on the right is in the Sun's inner corona.

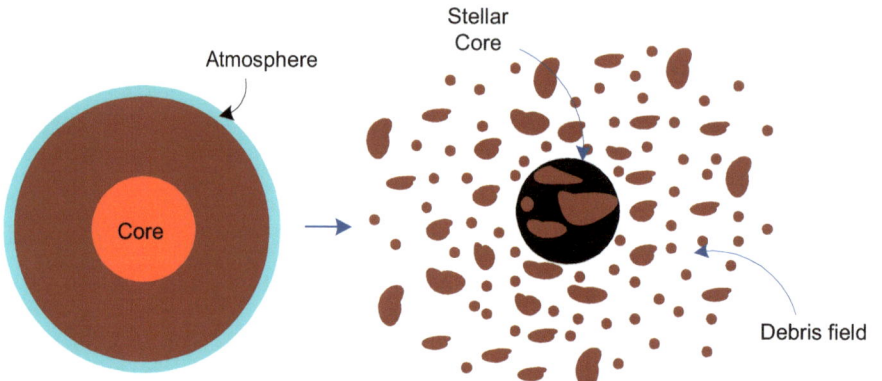

Figure 24.2. A living celestial object turns into a Stellar Core (SC) when it runs out of energy. A Stellar Core is made up of its original core and will be surrounded in a debris field. The debris field is the material that was once part of the object's outer layers of material. The water which used to be a part of the object will turn into tiny water droplets and surround the SC and the debris pieces that remain positively charged.

The Sun-like every other celestial object has internal and external energy levels. Energy decreases as we move outwards from the center. Objects of certain energy or at a certain energy level, as set by the Sun, moving in the Solar System can get to a minimum distance from the Sun, which corresponds to the position of the Sun's energy level corresponding to the object's own energy.

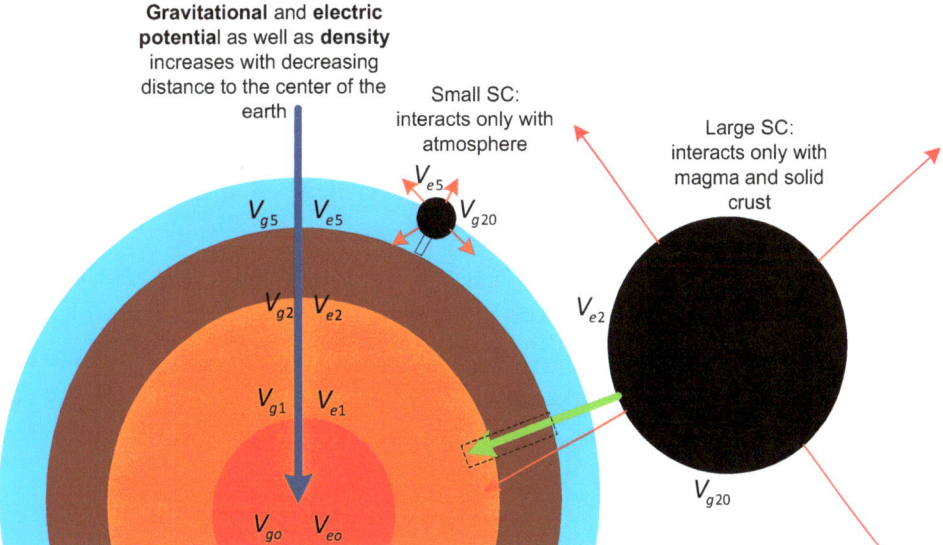

Figure 24.3. Celestial objects have internal energy levels which can be described in terms of their gravitational and electric potential: Stellar Cores (SCs) interact with matter, of the same proton density, it has, i.e. they interact with the matter layer at the same energy level as its own matter layers. So, if the SC has layers at different energy levels, matter at each energy level will interact with a layer, on the earth, at the same energy level.

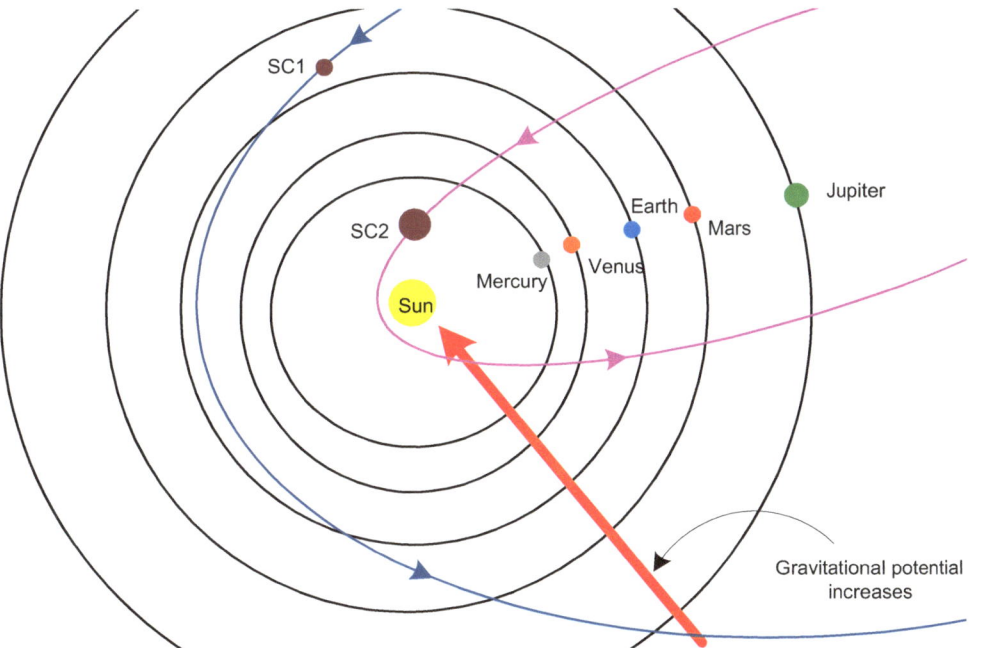

Figure 24.4. The Sun has external gravitational energy levels with associated decreasing gravitational potential as we move outwards from the Sun.

Now, large Stellar Cores seem to have material of a much higher energy level than earth and the smaller SCs seem to have material of a much lower energy level, to the point that they can only interact with earth's lower energy layers, which made me initially think that the earth would also have material at a certain energy levels and not others (see Article 569: Effect of Planet X Objects as large as the Sun on the Earth: energy levels) [2]. However, later I realized that the earth, as a living celestial object is different from a Stellar Core which is a dead celestial object and that the death process led to them losing matter at some of the energy levels that they initially had. The earth and all living celestial objects seem to have matter within which is at the maximum energy level possible which is matter that was originally a part of the super core from which all celestial objects in the universe come from (see Article 576: Planet X larger than the Sun on collision course with Earth: what happens?) [3]. This has to be the case so that a new planet can fully interact with its parent star and also stops any celestial object from colliding with any other celestial object no matter what the size. But asteroids are different, they are not celestial objects with cores in them, they are pieces of broken planets and they can impact the earth (see Article 541: Planet X cover-up and planetary formation: where do asteroids come from?) [4]

As these SCs approach the Sun surrounded by their debris fields, there will be many pieces of that debris field, which will not have the same gravitational energy, or attraction, for the Sun that the SC has and will thus be left behind somewhere in the Solar System. However, the pieces making up the inner part of the debris field will most likely proceed with the SC, which is why Saturn still has a debris field (see Article 559: Planet X, asteroid develops comet tail and alien ring maker) [5].

Figure 24.5. An artist's impression of Saturn (edge of uniform thickness and color indicates it is not a real photograph of Saturn), a re-energized Planet X System Stellar core which absorbed energy and material from the Sun a few thousand years ago, which turned it into a Solar System planet, a gas giant (see Article 523: Planet X and the Solar System: Jupiter and all gas giants are recent acquisitions) [6]. The rings are made out of what is left of its debris field and are kept in order by alien ring makers (see Article 559: Planet X, asteroid develops comet tail and alien ring maker) [5].

The pieces of debris will then go into orbit around the Sun at a distance from the Sun associated to their electrical and gravitational potentials. Since their gravitational potential will be much lower than their electrical potential they will attract material from the Sun's nebular clouds (fed by the Sun's solar wind) and will thus develop a tail, just like the Stellar Cores, until the debris pieces have absorbed enough electrons and gravitational potential to be neutral and have the same electric and gravitational potentials, at which time they will lose their cloud envelopes. This, therefore, explains where the asteroid belt came from; it was populated by the debris fields of Planet X System Stellar Cores going towards the Sun. The rocky pieces without a cloud envelope are most likely old pieces that have had time to absorb gravitational energy. Any asteroid that develops a tail is a newly arrived debris piece.

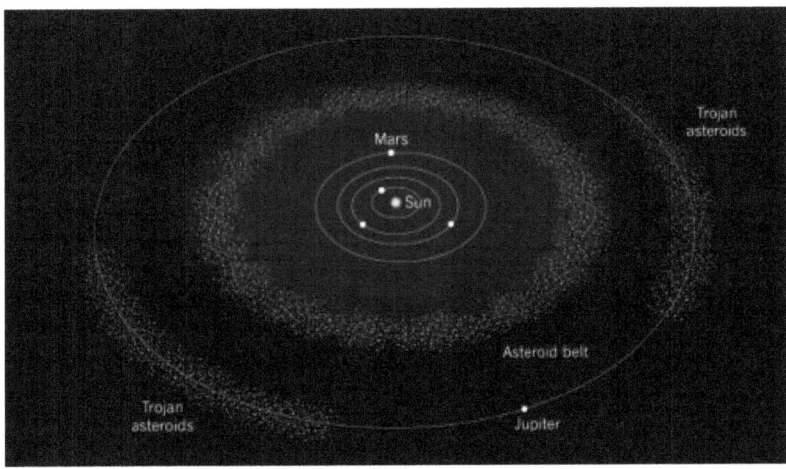

Figure 24.6. The asteroid belt, between Mars' and Jupiter's orbit and the asteroids in Jupiter's orbit, is Planet X debris left behind by the SC they were accompanied on their way to the Sun.

It is likely that every planet including earth has a similar asteroid belt as Jupiter, in its orbit. Some of this debris is likely to be a cloud, either separated from their rocky debris pieces or still attached to them if the objects are still absorbing energy and electrons. This is most likely where a large number of the earth's clouds are coming from, i.e. from debris pieces which have been left behind by large SCs on the way to the Sun, which have a similar electric potential as the earth but because they are so low in gravitational potential remain suspended in the atmosphere. Other clouds would come in with the Stellar Cores that the earth is hosting, i.e. they are coming into the earth's atmosphere to absorb energy from the earth.

Figure 24.7. A Planet X Object inside its cloud envelope, which does not completely cover it. The object is dark and clearly spherical. It is most likely a small moon core.

Figure 24.8. Left: The irregular and yet sharply angled shape of this dark cloud suggests that it is a large Planet X debris piece surrounded in a cloud envelope (Source: Secureteam 10 video: OMINOUS SPHERE: "It Just Sat There In The Fog.." from January 19th, 2019). **Right:** Sharp edges amongst the clouds indicate the presence of Planet X debris, i.e. pieces of broken up planets, suspended in the earth's atmosphere. These are low gravitational pieces of rock that most likely got deposited at the earth's orbit recently (Source: William Mignoli YouTube video).

This has not been the first time that rocks have been observed in the earth's atmosphere.

Figure 24.9. Dark shapes with jagged edges appear amongst these iridescent clouds. They look like mountain tops but that is clearly impossible as there is no other mountain behind the one in the foreground. There are no mountains to support the tops of what seem to be mountains. These must, therefore, be pieces of rock suspended in the atmosphere (see Article 546: Rocks suspended in the atmosphere amongst iridescent clouds) [7].

Asteroids that would come in and impact the earth are debris pieces that have been in the Solar System for a long time and have absorbed enough gravitational energy and electrons to now have higher gravitational energy than the earth's surface so that they fall toward the ground. It is likely that these objects maintain an interior positive charge like a Stellar Core but uniformly distributed and have an outer electron layer. Since they are charged, any attempt to land a probe on one will result in an electrical discharge, between the asteroid and the probe, as their negative electrical potentials equalize, i.e. electrons from the asteroid flow to the probe. This could be catastrophic for the probe.

In conclusion, asteroids are Planet X debris, which was deposited inside the Solar System, at some distance from the Sun, whilst the Stellar Core that they accompanied proceeded toward the Sun. They are likely to also be positively charged and have a cloud envelope, like the Stellar Cores, but will be uniformly charged rather than have the increasingly denser layers in the form of concentric shells that the SCs have. They will absorb electrons and gravitational energy inside the Solar System and from the earth if they enter its atmosphere; some asteroids will eventually be able to impact the earth's surface.

References:

[1] Albers, C. (2019). Article 575: Planet X causes atmospheric electron depletion, which is life-threatening.
[2] Albers, C. (2019). Article 569: Effect of Planet X Objects as large as the Sun on the Earth: energy levels.
[3] Albers, C. (2019). Article 576: Planet X larger than the Sun on a collision course with Earth: what happens?
[4] Albers, C. (2018). Article 541: Planet X cover-up and planetary formation: where do asteroids come from?
[5] Albers, C. (2019). Article 559: Planet X, asteroid develops comet tail and alien ring maker.
[6] Albers, C. (2018). Article 523: Planet X and the Solar System: Jupiter and all gas giants are recent acquisitions
[7] Albers, C. (2019). Article 546: Rocks suspended in the atmosphere amongst iridescent clouds.

The End for Now!

www.ingramcontent.com/pod-product-compliance
Lightning Source LLC
Chambersburg PA
CBHW051912210526
45473CB00006B/1988